A Puehler
8 December 2010
Bloomington, Indiana

Cultural Evolution

Contemporary Viewpoints

Cultural Evolution
Contemporary Viewpoints

Edited by

Gary M. Feinman

The Field Museum
Chicago, Illinois

and

Linda Manzanilla

Universidad Nacional Autonoma de Mexico
Mexico City, Mexico

Kluwer Academic / Plenum Publishers
New York Boston Dordrecht London Moscow

Library of Congress Cataloging-in-Publication Data

Cultural evolution: contemporary viewpoints/Gary M. Feinman and Linda Manzanilla.
p. cm.
Includes bibliographical references and index.
ISBN 0-306-46240-0
1. Human evolution. 2. Social evolution. 3. Archaeology—Methodology. 4. Indians—Antiquities. 5. Latin America—Antiquities. 6. Flannery, Kent V. I. Feinman, Gary M. II. Manzanilla, Linda.

GN281 .C85 2000
303.4—dc21 99-056491

ISBN: 0-306-46240-0

233 Spring Street, New York, N.Y. 10013

http://www.wkap.nl/

10 9 8 7 6 5 4 3 2 1

A C.I.P. record for this book is available from the Library of Congress

Printed in the United States of America

Kent V. Flannery

Contributors

Robert McC. Adams • Department of Anthropology, University of California–San Diego, La Jolla, California 92093

Robert L. Carneiro • Department of Anthropology, American Museum of Natural History, New York, New York 10024

Robert D. Drennan • Department of Anthropology, University of Pittsburgh, Pittsburgh, Pennsylvania 15260

Gary M. Feinman • Department of Anthropology, The Field Museum, Chicago, Illinois 60605

Marcella Frangipane • Dipartimento di Scienze Storiche Archeologiche e Antropologiche dell' Antichità, Università di Roma, 00185 Rome, Italy

Linda Manzanilla • Instituto de Investigaciones Antropológicas, Universidad Nacional Autónoma de México, Ciudad Universitaria, 04510 Mexico City, Mexico

Linda M. Nicholas • Department of Anthropology, The Field Museum, Chicago, Illinois 60605

Elsa M. Redmond • Department of Anthropology, American Museum of Natural History, New York, New York 10024

Bruce D. Smith • Department of Anthropology, National Museum of Natural History, Smithsonian Institution, Washington, D.C. 20560

Charles S. Spencer • Department of Anthropology, American Museum of Natural History, New York, New York 10024

Henry D. Wright • Museum of Anthropology, University of Michigan, Ann Arbor, Michigan 48109

Preface

The chapters in this volume express the current diversity and richness of contemporary cultural evolutionary approaches. Although most of the chapters represent specific case studies, each of the authors structures his or her analysis with broader comparative issues in central focus. Collectively, these papers express the richness of the issues that are under investigation by comparative theorists interested in long-term change as well as the present diversity of data, approaches, and ideas that are being studied and employed to examine these questions.

Each contributor also recognizes the key role that Kent V. Flannery has played over the past forty years in the development of such cross-cultural approaches. His ideas about the origins of agriculture, economic change, and the development of hierarchical political organizations have helped lay a foundation for many of the approaches taken in this volume. For that reason, all of those involved (each of us students, colleagues, and/or close associates of Flannery) wish to dedicate this collection to him.

As with many edited collections, we have traveled a bit of a winding road toward publication. For this reason, Linda Manzanilla and I, as editors, gratefully acknowledge all the people who helped make this volume a reality. We wish to thank Eliot Werner, our editor at Kluwer Academic/Plenum Publishers, for his great support and help during all stages of this project. We also owe a large debt to Linda M. Nicholas, whose editorial and organizational skills facilitated our work in countless ways. Joyce Marcus provided tremendous assistance both by compiling Flannery's bibliography and by providing us with the photograph that appears as the frontispiece. The editors also wish to recognize the authors for their thoughtful contributions, as well as for their understanding and willingness to cooperate with us throughout the production process.

As senior editor, it gives me great pleasure to write this preface just kilometers from the mouth of Guilá Naquitz cave as well as other sites that have proved so important to world archaeology, the study of cultural evolution, and the contributions and career of "Señor Kent" (as many down here still know him). May this volume serve as a valued contribution to the comparative investigation of long-term change and a fitting tribute to the Master.

GARY M. FEINMAN
Mitla, Oaxaca, Mexico
March 1999

Contents

Part I

Introduction

Chapter **1**

Cultural Evolutionary Approaches and Archaeology

Past, Present, and Future

GARY M. FEINMAN

INTRODUCTION

A dedicated archaeologist returning to the local state university library after two isolated decades in the field would find himself rather surprised at and befuddled by the theoretical state of the discipline, especially if he was asked to prepare a course on archaeology's contemporary method and theory. The returning archaeologist would find that broadly defined processual or sociocultural evolutionary approaches that had predominated in the late 1970s when he or she left for the field are still central to much current archaeological practice (Schiffer, 1996:643). Although such contemporary perspectives have not been static, many of the overarching research questions about long-term change (such as the rise of inequality and states in different regions and the emergence of agriculture) remain critical concerns as they have been for decades. Yet a stunning paradox is obvious: In a growing list of purely theoretical, "armchair" expositions, the long-away field-worker would find that select critics use a slew of invectives to rebuke those same predominating theoretical approaches. Depending on who is read, the returnee would find that the study of cultural evolution is described (e.g., Lyman

GARY M. FEINMAN • Department of Anthropology, The Field Museum, Chicago, Illinois 60605

Cultural Evolution: Contemporary Viewpoints, edited by Gary M. Feinman and Linda Manzanilla. Kluwer Academic/Plenum Publishers, 2000.

and O'Brien, 1996; O'Brien, 1996; Shanks and Tilley, 1987) as inherently "reductionist," "vitalist," "essentialist," "progressivist," and even "ethnocentric."

In this volume (and the present introduction), the goal is not to mount a direct defense of cultural evolutionary approaches against the aforementioned castigations. General discussions that offer a range of responses to such challenges have recently been published elsewhere (e.g., Blanton et al., 1993:10-23; Carneiro, 1996; Flannery, 1995; Rambo, 1991; Rambo et al., 1991; Sanderson, 1990; Spencer, 1997). Rather, the aims of this collective work are threefold. The first is to illustrate through case studies and theoretical essays the vitality of contemporary cultural evolutionary approaches, both in terms of the collection and interpretation of new data and the definition and refinement of specific new research questions. The second aim, in accord with Spencer (1997), is to exemplify the diversity of such current cultural evolutionary approaches and to highlight key areas of ongoing debate among its practitioners. As both critics and proponents of this broad framework have recognized (e.g., Dunnell, 1980; Sanderson, 1990; Spencer, 1997), cultural evolutionary approaches in archaeology have themselves shifted greatly over the last century or so, although at the most general level major research queries have been somewhat more stable. The intellectual centrality of these overarching questions, the productivity of the approach, and the vibrancy of current discussion among cultural evolutionary theorists seem to justify this framework's predominant role in contemporary archaeological practice. A final (yet key) goal of the volume is to accent the focal role of the contributions of Kent V. Flannery to current cultural evolutionary research.

The remainder of this brief introductory essay identifies what I mean by a cultural evolutionary approach and discusses some aspects of the past, present, and future of this perspective in relation to general archaeological research. By so doing, this introduction also serves as a vehicle to outline the chapters that compose the rest of this collection.

EVOLUTIONARY APPROACHES

The study of cultural evolution has been a central concern of anthropological archaeology for more than one century (see Sanderson, 1990). Throughout these years, authors have proposed and employed diverse approaches. Many current criticisms of cultural evolution are more applicable to the works of Lewis Henry Morgan or Leslie White than they are to present practice. Consequently, it is necessary to define what is meant here by a broad cultural evolutionary approach and to provide a degree of historical background for such contemporary theoretical frameworks. Nevertheless, I preface my discussion with a few words about evolutionary approaches in general.

In recent decades most, although perhaps not all, theorists have recognized that biological and cultural evolution are distinct theories used to understand different phenomena. For example, Stephen J. Gould (1987:70) has stated that "fruitful analogies may be drawn between biological and cultural evolution, but they remain analogies. The processes are different, even though human culture has a biological base. Cultural evolution needs laws of its own." Likewise, Flannery (1995:3) has recognized that "biological evolution is an imperfect analogy for social evolution." In certain ways the analogy is useful, and in other ways it can mislead.

Of course, both biological and social evolution are concerned with complex episodes of long-term change. Perhaps as a consequence, both theoretical concepts have two distinct but related meanings (Rambo, 1991:26). In one sense, the term *evolution* refers to a temporal sequence or succession of forms, for example, the evolution of the horse. When biologists claim that evolution is a fact rather than a theory, such evolutionary sequences and replacements through time are their evidential basis. The second meaning of *evolution* is that of the causal process by which change in forms (or populations) occurs over time. In regard to biology, Darwin was most concerned with this second meaning—the understanding of biological evolutionary process. Nevertheless, the present debates in biology over the tempo of evolutionary change (punctuated equilibrium versus gradualism) are simultaneously concerned with sequence and process: "This continuous interplay between improved knowledge of sequences and more powerful models of process contributes to the continued vigor of evolutionary studies in biology" (Rambo, 1991:27).

I have reviewed these dual meanings of evolution because a number of theorists have proposed that the differential emphasis between them represents a principal distinction between biological and cultural evolutionary thought. For example, Rambo (1991:27) opined that, in contrast to Darwinian evolution, "reconstruction of sequence dominates cultural evolutionary studies" in the social sciences. He went on to suggest that in social evolutionary research there is almost no integration of research on sequence and process. More pointedly, the selectionist critiques (e.g., O'Brien, 1996) of cultural evolutionary studies as "essentialist" presume that such studies are concerned entirely with a fixed sequence of societal forms, types, or "essences." Here it is argued that, whereas certain variants of cultural evolutionary theory were indeed overly concerned with sequence as opposed to process, this characterization is inapplicable to many current studies or concepts of cultural evolution. In fact, at least as early as 1972 (Flannery, 1972; see also Drennan, 1991), archaeologists began to devote serious consideration to process as well as sequence. The interplay between the two (process and sequence) lies behind the growing concern with agency (e.g., Blanton et al., 1996; Brumfiel, 1994; Spencer, 1993, 1997) and "action theory" (Marcus and Flannery, 1996) in contemporary cultural evolutionary thought. Although no agreement on understanding process has yet been reached, this fertile area of ongoing discussion requires much further consideration (see Drennan, Chap. 7, this volume).

DEFINING A CULTURAL EVOLUTIONARY FRAMEWORK

Questions about the long-term patterns of social change and the historical causes of cultural variation are older than either the concept of cultural evolution or the academic discipline of anthropology (e.g., Harris, 1968). In fact, the birth of the field is linked with fundamental objectives about the historical development of specific hierarchical political institutions and basic technologies, such as metallurgy (e.g., Tyler, 1960:11–13). If one takes into account changes in the terminological jargon, comparable general research questions remain at the core of contemporary cultural evolution (although the underlying assumptions, the empirical record, and the theories of process clearly have undergone dramatic transformations).

Recently, for example, Carneiro (1996:271) outlined the domain of cultural evolution as encompassing the investigation of the processes that account for the transformations from the social relations and technologies of the Paleolithic to the complex states of today. I believe that this definition encompasses the research aims and objectives of most contemporary students of cultural evolution. Yet even among its practitioners, there is no agreement on exactly what constitutes a cultural evolutionary framework. Several contemporary sociologists tend to equate evolutionary theories first and foremost with a typology of social forms (Sanderson, 1990; Wright, 1983). In other words, sequence is given precedence over process in these views as the selectionist critiques also contend (Leonard and Jones, 1987:202; O'Brien, 1996). Yet this definition of cultural evolutionary approaches is too narrow and does not correspond with much recent practice in anthropological archaeology. For example, my colleagues and I (Blanton et al., 1993:10) have written, "By saying that an evolutionary approach seems like the best direction for anthropology, we mean something other than the construction of general sequences of progressive stages." Likewise Drennan (1991:114) correspondingly stated, "For me, the essence of cultural evolution is not directionality of change through a proscribed set of stages but the way processes are modeled. These processes, in a cultural evolutionary model, must include mechanisms to generate variability and mechanisms to select from that variability the patterns that are perpetuated."

CONTEMPORARY CULTURAL EVOLUTIONARY APPROACHES: WHAT THEY ARE AND ARE NOT

Although the domain of study for cultural evolution was first broadly defined during the last century, basic assumptions and interpretations about cultural diversity and change have undergone dramatic revision over more than one hundred years. Consequently, it is unproductive and inappropriate to critique contemporary cultural evolutionary perspectives for the mistakes, limitations, and omissions of past perspectives that were advanced many years ago. As noted above, nineteenth-century cultural evolutionary thought was more concerned with typological stages and sequence than with process (Rambo, 1991:29; Spencer, 1997:212). The empirical basis for these sequences was largely synchronic cases drawn from ethnographic and historic data, which were sometimes ordered so that contemporary Western societies were placed at the pinnacle of the sequential ladder. The postulated trajectories of evolutionary transition were equated with progress, so that long-term historical change was perceived as goal directed or moving in a preordained positive direction (e.g., toward greater efficiency, problem-solving ability, or some morally defined position) (Granovetter, 1979; Voget, 1975:195-198). The engine of change was often determined to be biological, so that major advances in human physical capabilities (mental abilities) were viewed as responsible for cultural transitions.

However, it is wrong to presume that a comparative interest in long-term cultural change, or even the recognition of certain diachronic directional trends that have characterized the course of human history, necessarily implies an acceptance of the notion of progress (either in the sense of value judgments or perceived directionality toward a preordained goal). For example, there is no question that human sociopolitical formations generally have become larger and more hierarchically organized over the last tens of thousands of years. However, this observation does not entail the judgment that such formations are intrinsically better, more efficient, or more advanced. Neither does it mean that such directional trends necessarily continue unabated through time, or that growth or change in a particular direction always occurs in a specific historical context. The course of cultural change has been complex and contingent (Feinman, 1994:19), not unidirectional (Carneiro, 1996:273).

Modern cultural evolutionary thought also has long rejected the nineteenth-century notion that shifting biological capabilities alone have caused episodes of cultural transition (Harris, 1968:636; White, 1949:141). Likewise, over the last century, the study of cultural evolution has shifted from gathering an empirical record of largely synchronic cases that were presumed to represent a uniform pathway of change to compiling a global archaeological database that permits actual diachronic examinations of different patterns and sequences of change in many world regions. This shift in empirical emphasis is important because it has allowed theorists to see that cultural change generally does not occur in a regular or necessarily steplike fashion. Although directional trends (not laws) are present (and require eventual explanation), the marked variation from region to region in the tempo and patterns of specific sequences of long-term change directs our focus away from studying cultures or societies in their entirety. For that reason, Flannery (1995) has suggested that we call this present theoretical stream of research social—as opposed to cultural—evolution.

As Spencer (1997) has noted, debates in contemporary cultural evolutionary studies extend well beyond how to refer to this research domain. Some researchers give somewhat greater emphasis to sequences of change (e.g., Flannery, 1995; Spencer, 1997); others stress the study of process (Drennan, 1991). There also is a healthy debate over the use of evolutionary stages, levels, and recurring designs (or *Baupläne*) (Blanton et al., 1993; Feinman and Neitzel, 1984; Sanderson, 1990; Spencer, 1997:235; Upham, 1987, 1990). Whereas some scholars emphasize cross-cultural commonalities in the form of specific societal institutions and the ways that changes in one institution co-occur with shifts in others, other theorists see societal change as less steplike. These investigators reject the notion that from one historical case to another key shifts in one societal institution necessarily entail comparable shifts in others. In other words, change in different aspects of society may be decoupled, and more variable sequences of long-term change might be expected. Once again, if we are to evaluate these alternatives, we cannot always study the evolution of cultures in their entirety, and we must examine the nature and tempo of change in different facets of cultural and social behavior and organization.

These debates largely concern matters of emphasis in addition to expectations that will ultimately be better investigated and resolved through the systematic collection of additional archaeological and historical data. For that reason, these ongoing discussions and debates are fruitful, continuing a century of refinement in cultural evolutionary thought. Earlier debates among cultural evolutionists, such as White, Steward, Service, Fried, Binford, and Flannery, indeed have helped generate the empirical record relevant to social and cultural evolutionary issues that we study today.

CONTRIBUTIONS TO CONTEMPORARY CULTURAL EVOLUTION IN THIS VOLUME

As Gould (1988:333) has written, "What the science of history requires above all is a tractable way to study directional processes." As noted above, such studies may search for and expect evidence of steplike transitions or changes toward greater technological efficiency, or they may use a more multilinear framework that decouples and investigates the nature of the changes in the relation between different institutions and/or patterns of behavior. The latter approach often concentrates on specific institutional or technological shifts that seem to have analogous parallels in different historical settings (Haas, 1997). Such shifts may include (but are by no means limited to) the rise of state institutions, the initial aggregation in cities, the development of sociopolitical hierarchies, the emergence of domestication, and the invention of new technologies.

In Part II of this volume, four chapters (Smith, Carneiro, Adams, and Feinman and Nicholas) examine ancient technologies and episodes of technological change.

Smith presents new chronometric and botanical evidence relevant to Flannery's (1986) claims for early plant domestication at Guilá Naquitz cave in the eastern arm of the Valley of Oaxaca, Mexico. Smith's new findings support Flannery's original interpretations, which solidly place the domestication of cucurbits in southern Mexico thousands of years before sedentary life was adopted in this part of the world. Smith's confirmation of Flannery's earlier interpretations is important because other claims for early agriculture in Mexico have been challenged (Long et al., 1989) and because it indicates that the timing and sequence of the shifts leading to farming and sedentary life in highland Mexico were distinct from those in other global regions, for example, Southwest Asia.

The contributions by Carneiro and Adams offer rather different perspectives on episodes of technological change. In his thorough examination of the *tipití* and manioc-processing technologies in lowland South America, Carneiro endeavors to illustrate that the historical sequence of innovation followed a relatively orderly pattern toward increased energetic efficiency. Through his discussion of the *tipití* and its proposed antecedents, Carneiro aims to provide general support for the cultural evolutionary paradigm championed by Leslie White (1949) a half-century ago. Adams takes an alternative vantage in his discussion of major technological breakthroughs in Southwest Asia. He sees the course of technological innovation as less direct or predictable. Adams also eschews a steplike view of change in his argument that major technological innovations cannot necessarily be coupled with (or explained by) major sociopolitical shifts.

In the final chapter in Part II, Feinman and Nicholas argue against a unilinear model of specialized production through a discussion of excavated materials that they recently collected at the Ejutla site in the Southern Highlands of Mexico. Feinman and Nicholas argue that the inhabitants of a Classic-period (A.D. 200–800) Ejutla domestic residence were specialists, in that they produced shell ornaments for exchange (and probably distant exchanges), but they worked in a domestic context. The findings thereby question the unilinear view that specialized production ought to entail nondomestic workshop manufacture.

Part III also includes four chapters (Spencer and Redmond, Drennan, Wright, and Frangipane) that are followed by a brief conclusion, a postscript, and a full list of the publications of Kent Flannery. These chapters focus primarily on sociopolitical change, including both a case study and several theoretical explorations. Spencer and Redmond holistically examine political, economic, and religious changes during the Middle-Terminal Formative period (600 B.C.–A.D. 200) in the Cuicatlán Cañada (Oaxaca, Mexico). They argue that Monte Albán's conquest of the Cañada (located in the Valley of Oaxaca) led to marked changes in most spheres of life in the conquered region. Their holistic approach to major episodes of cultural change owes much to views long championed by Flannery (1972; Marcus and Flannery, 1996).

The final three chapters are more explicitly theoretical efforts that consider the relationships between leaders and followers and/or centers and outlying settlements. Such relationships lie at the heart of past and present inquiries about different strategies of socioeconomic integration and political control. Drennan endeavors to unpack the relation between structure and agency in an effort to build a sophisticated approach to process and diversity in contemporary cultural evolutionary thought. Both Wright and Frangipane explore the different socioeconomic linkages that can tie smaller communities to larger ones. Wright's analysis builds on an earlier classic paper by Steponaitis (1978) that modeled the possible effects of tribute payments on settlement patterns in chiefdoms. Yet material tribute is not universal to chiefly societies, and Wright endeavors to advance an alternative settlement model for cases in which dependence relations were more apt to be paid in labor than in goods. Although still preliminary, Wright's labor model defines an important new avenue to understand settlement pattern variation in different chiefly contexts. Focused generally on Southwest Asia, Frangipane looks at the different patterns and strategies of centralized exchange. By so doing, she works toward a more precise definition of redistribution, one that highlights key temporal shifts in the nature of centralized patterns of exchange throughout the ancient Near East.

CONCLUDING THOUGHTS

To understand the long sweep of human history, we still must come to understand how and why major sociopolitical, technological, and economic changes—such as the rise of state forms of government and the beginnings of agriculture—occurred in different regions of the world. It is important to know how similar were the processes that led to change and how comparable were the historical sequences in different areas. These issues are as central to our understanding of humankind as they were in the past. Although these questions need not be formulated in a typological fashion, coming to grips with these queries is basic to our knowledge of global histories and long-term change.

Important theoretical and enormous empirical advances have occurred over the last century, but the major social shifts in human institutions and technologies that have been posed are difficult, complex issues. Many additional generations of research and theorizing may be necessary before fully satisfying and convincing answers are achieved. Archaeology plays a central role in addressing the broad suite of issues highlighted by cultural evolutionary thought. Consequently, these core topics must continue to serve as a general guide for archaeological investigation, as they have for at least the last several decades.

Yet as discussed above, although cultural evolutionary issues have had a central role in archaeology for years, the approach itself has been far from static. The framework continues to be modified and challenged by practitioners and critics. The vitality of the currently debated issues and the centrality of the questions under investigation provide ample evidence that a broad cultural evolutionary framework remains vibrant despite the slew of invectives that have been hurled in its direction. With its comparativist roots along with increasing concern with process and variability, a broad cultural evolutionary perspective still provides a thoughtful intellectual roadmap for twenty-first-century archaeology. If that roadmap is followed by forthcoming generations of archaeologists, then the twentieth-century writings of Kent Flannery will have had a central and essential role in establishing the intellectual underpinnings for those future directions.

REFERENCES

Blanton, R. E., Feinman, G. M., Kowalewski, S. A., and Peregrine, P. N., 1996, A Dual-Processual Theory for the Evolution of Mesoamerican Civilization. *Current Anthropology* 37:1–14.

Blanton, R. E., Kowalewski, S. A, Feinman, G. M., and Finsten, L. M., 1993, *Ancient Mesoamerica: A Comparison of Change in Three Regions*. 2nd ed. Cambridge University Press, Cambridge.

Brumfiel, E. M., 1994, Factional Competition and Political Development in the New World: An Introduction. In *Factional Competition and Political Development in the New World*, edited by E. M. Brumfiel and J. W. Fox, pp. 3–13. Cambridge University Press, Cambridge.

Carneiro, R. L., 1996, Cultural Evolution. In *Encyclopedia of Cultural Anthropology*, edited by D. Levinson and M. Ember, pp. 271–277. Henry Holt, New York.

Drennan, R. D., 1991, Cultural Evolution, Human Ecology, and Empirical Research. In *Profiles in Cultural Evolution: Papers from a Conference in Honor of Elman R. Service*, edited by A. T. Rambo and K. Gillogly, pp. 113–135. Anthropological Papers No. 85. Museum of Anthropology, University of Michigan, Ann Arbor.

Dunnell, R. C., 1980, Evolutionary Theory and Archaeology. In *Advances in Archaeological Method and Theory*, Volume 3, edited by M. B. Schiffer, pp. 35–99. Academic Press, New York.

Feinman, G. M., 1994, Toward an Archaeology without Polarization: Comments on Contemporary Theory. In *Caciques and Their People: A Volume in Honor of Ronald Spores*, edited by J. Marcus and J. F. Zeitlin, pp. 13–43. Anthropological Papers No. 89. Museum of Anthropology, University of Michigan, Ann Arbor.

Feinman, G. M., and Neitzel, J., 1984, Too Many Types: An Overview of Sedentary Prestate Societies in the Americas. In *Advances in Archaeological Method and Theory*, Volume 7, edited by M.B. Schiffer, pp. 39–102. Academic Press, New York.

Flannery, K. V., 1972, The Cultural Evolution of Civilizations. *Annual Review of Ecology and Systematics* 3:399–426.

Flannery, K. V., 1995, Prehistoric Social Evolution. In *Research Frontiers in Anthropology*, edited by C. R. Ember and M. Ember, pp. 1–26. Prentice-Hall, Englewood Cliffs, NJ.

Flannery, K. V. (editor), 1986, *Guilá Naquitz: Archaic Foraging and Early Agriculture in Oaxaca, Mexico*. Academic Press, New York.

Gould, S. J., 1987, The Ghost of Patagoras. In *An Urchin in the Storm: Essays about Books and Ideas*, by S. J. Gould, pp. 62–72. Norton, New York.

Gould, S. J., 1988, On Replacing the Idea of Progress with an Operational Notion of Directionality. In *Evolutionary Progress*, edited by M. H. Nitecki, pp. 319–338. University of Chicago Press, Chicago.

Granovetter, M., 1979, The Idea of "Advancement" in Theories of Social Evolution and Development. *American Journal of Sociology* 85:489–515.

Haas, J., 1997, A Brief Consideration of Cultural Evolution: States, Agents and Tinkering. Unpublished manuscript, Department of Anthopology, The Field Museum, Chicago.

Harris, M., 1968, *The Rise of Anthropological Theory*. Crowell, New York.
Leonard, R. D., and Jones, G. T., 1987, Elements of an Inclusive Evolutionary Model for Archaeology. *Journal of Anthropological Archaeology* 6:199–219.
Long, A., Benz, B., Donahue, J., Hull, A., and Toolin, J., 1989, First Direct AMS Dates on Early Maize from Tehuacan, Mexico. *Radiocarbon* 31:1035–1040.
Lyman, R. L., and O'Brien, M. J., 1996, The Concept of Evolution in Early Twentieth Century Americanist Archeology. In *Rediscovering Darwin: Evolutionary Theory in Archeological Explanation*, edited by C. M. Barton and G. A. Clark, pp. 21–48. Archeological Papers No. 7. American Anthropological Association, Washington, D.C.
Marcus, J., and Flannery, K. V., 1996, *Zapotec Civilization: How Urban Society Evolved in Mexico's Oaxaca Valley*. Thames and Hudson, London.
O'Brien, M. J., 1996, Evolutionary Archaeology: An Introduction. In *Evolutionary Archaeology: Theory and Application*, edited by M. J. O'Brien and R. C. Dunnell, pp. 1–15. University of Utah Press, Salt Lake City.
Rambo, A. T., 1991, The Study of Cultural Evolution. In *Profiles in Cultural Evolution: Papers from a Conference in Honor of Elman R. Service*, edited by A. T. Rambo and K. Gillogly, pp. 23–109. Anthropological Papers No. 85. Museum of Anthropology, University of Michigan, Ann Arbor.
Rambo, A. T., Gillogly, K., Hutterer, K. L., and Parsons, J. R., 1991, Introduction. In *Profiles in Cultural Evolution: Papers from a Conference in Honor of Elman R. Service*, edited by A. T. Rambo and K. Gillogly, pp. xiii–xvi. Anthropological Papers No. 85. Museum of Anthropology, University of Michigan, Ann Arbor.
Sanderson, S. K., 1990, *Social Evolutionism: A Critical History*. Blackwell, Cambridge, MA.
Schiffer, M. B. 1996. Some Relationships between Behavioral and Evolutionary Archaeologies. *American Antiquity* 61:643–662.
Shanks, M., and Tilley, C., 1987, *Social Theory and Archaeology*. University of New Mexico Press, Albuquerque.
Spencer, C. S., 1993, Human Agency, Biased Transmission, and the Cultural Evolution of Chiefly Authority. *Journal of Anthropological Archaeology* 12:41–74.
Spencer, C. S., 1997, Evolutionary Approaches in Archaeology. *Journal of Archaeological Research* 5:209–264.
Steponaitis, V. P., 1978, Location Theory and Complex Chiefdoms: A Mississippian Example. In *Mississippian Settlement Patterns*, edited by B. Smith, pp. 417–453. Academic Press, New York.
Tylor, E. B., 1960, *Anthropology*. University of Michigan Press, Ann Arbor [originally published 1881].
Upham, S., 1987, A Theoretical Consideration of Middle Range Societies. In *Chiefdoms in the Americas*, edited by R. D. Drennan and C. A. Uribe, pp. 345–367. University Press of America, Lanham, MD.
Upham, S., 1990, Decoupling the Process of Political Evolution. In *The Evolution of Political Systems: Sociopolitics in Small-Scale Sedentary Societies*, edited by S. Upham, pp. 1–17. Cambridge University Press, Cambridge.
Voget, F. W., 1975, *A History of Ethnology*. Holt, Rinehart and Winston, New York.
White, L. A., 1949, *The Science of Culture: A Study of Man and Civilization*. Grove Press, New York.
Wright, E. O., 1983, Giddens's Critique of Marxism. *New Left Review* 138:11–35.

Part II

Technological Change and Economic Specialization

Although diverse, the four chapters in this section share a focus on technology, economic change, and division of labor. The specific case studies and historical contexts vary greatly, yet each speaks to ongoing theoretical debates about the nature and uniformity or diversity of technological change processes throughout the course of human history. Whereas Carneiro argues for a rather direct and efficiency-driven perspective guided by his study of the *tipití* in lowland South America, Adams takes a broader and more complex view of technological change during the course of Near Eastern history. As expressed here, Adams's perspective seems to afford a far greater role to human organization and agency than does Carneiro's, and hence places less emphasis on efficiency or skill alone. Nevertheless, both authors neatly weave a consideration of objects and archaeological findings with historical (Adams) and ethnographic (Carneiro) sources.

By implication, the chapters by Smith and by Feinman and Nicholas also emphasize variation in the course of technological and economic change in distinct global regions. Smith's careful and significant reanalysis of Kent Flannery's early and important findings of domestic squash at Guilá Naquitz cave supports Flannery's earlier interpretations. Together, these findings strengthen the view that the process and cultural implications of plant domestication in highland Mexico may have been rather different from the course of early grass (wheat and barley) domestication in the Near East. For example, whereas plant domestication in highland Mexico (and at Guilá Naquitz) still appears to have occurred among low-density mobile populations, sedentism appears to precede plant domestication in the Near East.

Likewise, Feinman and Nicholas's interpretation of household-based craft specialization (shell working) in ancient Ejutla does not conform to the contemporary Euro-American (and often broadly assumed) notion that serious work (such as specialized craft production for exchange) takes place in workshop (nondomestic) contexts. By implication, this perspective implies that the nature of work, specialization, and exchange and distribution in ancient Mesoamerica may have differed from other economies (such as the Roman world) where workshops were relatively common and high-intensity production (by volume) generally cooccurred with large-scale (nondomestic) production.

Each chapter in this section illustrates important parallels with the works of Flannery. All authors take an explicitly comparative focus (albeit at different analytical scales). In so doing, significant historical questions about long-term processes of change underlie each chapter. In addition, each also integrates multiple and diverse strands of evidence to reach its conclusions. Collectively, these chapters make use of botanical, zoological, documentary, and ethnographic data in a manner that best exemplifies the holistic investigations that Flannery has championed throughout his career.

Chapter 2

Guilá Naquitz Revisited

Agricultural Origins in Oaxaca, Mexico

BRUCE D. SMITH

INTRODUCTION: A SMALL AND SHALLOW CAVE

In 1964, with Richard S. MacNeish's Tehuacán project drawing to a close, Kent Flannery selected the Valley of Oaxaca, located 150 km to the south, as a promising region in which to continue the search for evidence of the developmental transition from foraging to farming in Mesoamerica. Over the next two years, more than 60 caves and rock shelters that might contain evidence of early plant domestication were located in initial surveys that focused on an area of the Oaxaca Valley northwest of Mitla (Figure 1). Six of these caves were situated along Guilá Naquitz ("White Cliff"), which extends for more than 800 m along the northern edge of a small tributary canyon of the Río Mitla. One of these caves, a small 8-by-10-m opening at the base of the volcanic tuff canyon wall, was given the name Guilá Naquitz, and, because of its apparent preceramic-period occupations and its promise of well-preserved plant remains, it was selected for full-scale investigation (Figure 2).

When excavated in April and May 1966, Guilá Naquitz was found to contain a 20-cm-thick layer (Zone A) dating to c. A.D. 620–740 (Monte Albán IIIb–IV) and, below this, six preceramic layers (Zones B1, B2, B3, C, D, E)[1] resulting from a series of intermittent short-term seasonal occupations by small microband family groups that dated between about 8900 and 6700 B.C. in radiocarbon years. This

BRUCE D. SMITH • Department of Anthropology, National Museum of Natural History, Smithsonian Institution, Washington, D.C. 20560.

Cultural Evolution: Contemporary Viewpoints, edited by Gary M. Feinman and Linda Manzanilla. Kluwer Academic/Plenum Publishers, 2000.

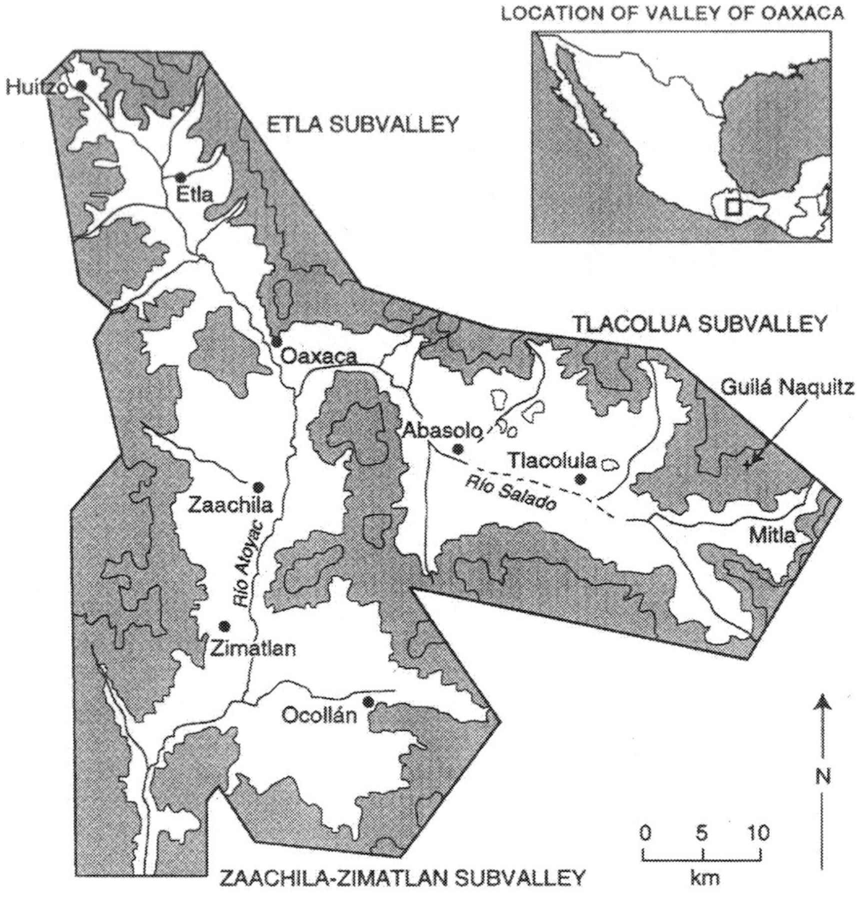

Figure 1. The location of Guilá Naquitz cave in the Valley of Oaxaca (after Flannery 1986:Fig. 3.1).

record of intermittent seasonal use of Guilá Naquitz by Archaic-period groups appeared to span the transition from hunting and gathering to initial food production in Oaxaca. Scattered in among the animal bones, wild plant remains, artifacts, and hearths of Zones B1–E were a few rind fragments of bottle gourd (*Lagenaria siceraria*), along with the seeds, peduncles (fruit stems), and rind fragments of an apparent domesticated squash (*Cucurbita pepo*).

More than 30 years after its discovery, this apparent early domesticated squash, and its presence in the Archaic-period zones of Guilá Naquitz, continues to occupy a critically important central role in our current understanding of the origins

Figure 2. View of Guilá Naquitz cave from the east in 1999. The dark opening of the cave (arrow) is just visible behind vegetation that has grown up in the 30 years since it was excavated (photo by Linda Nicholas).

of agriculture in Mesoamerica. The most compelling explanatory framework for the initial domestication of plants and the transition to food production in Mexico appears in the landmark volume *Guilá Naquitz: Archaic Foraging and Early Agriculture in Oaxaca, Mexico* (Flannery, 1986). This book can certainly be characterized as a site report, because it presents a detailed description of both the excavation of Guilá Naquitz and the materials recovered and an analysis and interpretation of the range of human activities carried out by the Archaic-period inhabitants of the cave. In addition to considering the record of human occupation of Guilá Naquitz, however, Flannery also placed the preceramic societies of the Oaxaca Valley in a broad ecological context and viewed them from a developmental perspective as he addressed the processual questions of when, how, and why the population transformed an indigenous wild *Cucurbita* gourd into a domesticated crop plant—a *Cucurbita pepo* squash. This processual explanation for the initial development of food production in Oaxaca, which rests in large measure on the excavation of Guilá Naquitz and the initial appearance there of domesticated squash, has in turn formed the basis for more abstract interpretive models of agricultural origins for Mesoamerica and other regions of the world (Smith, 1995a, 1998). In addition, Flannery also carried the discussion of agricultural origins to an even higher plane in the Guilá Naquitz volume, as he "visits the master" and contemplated the general epistemological basis for causal explanation of human cultural evolution in general and the transition to food production in particular.

The Guilá Naquitz volume, then, considered the origins of agriculture on a number of different levels. In the process, Flannery built upward and outward from Guilá Naquitz and its apparently domesticated squash to include (1) a site-level analysis of the range of activities carried out by the cave's microband family group; (2) a settlement-system-level consideration of Guilá Naquitz cave in the annual cycle/seasonal round of its larger macroband multifamily unit; (3) a valley-level interpretation of the unfolding temporal trajectory of preceramic Oaxacan societies' adjustment and response to environmental constraints and opportunities as they made the initial transition to food production; and finally, (4) a contemplation of the appropriate general explanatory framework in which to address such complex episodes of cultural transformation.

In the decade and more that has passed since its publication, the Guilá Naquitz volume has understandably received considerable attention and accolades for the clarity and quality of discourse on all its ascending levels of analysis. A few scholars, however, have raised concerns about the relative strength of the detailed arguments that connect and support the different spatial, temporal, and topical levels of description and analysis. Dumond, for example, concluded his one-page review of the Guilá Naquitz volume with a general expression of doubt about the link between higher level interpretation and primary evidence of human activities uncovered in Guilá Naquitz cave: "Thus, one is pushed to an uneasy if familiar position: the simulation and related arguments regarding the evolution of subsistence techniques are attractive, even compelling. But do they touch reality?" (Dumond, 1987:195). Although Dumond remarked on the small size and shallow depth of deposits in Guilá Naquitz and its "paltry collection of artifacts" (Dumond, 1987:195), the primary focus of his skepticism was what he saw as the questionable stratigraphic integrity of the cave's preceramic habitation zones and the associated problem of the true temporal placement of the bottle gourd and domesticated *Cucurbita pepo* squash specimens:

> One can still imagine early charcoal in later floors, some of which (perhaps all of zone B?) date after 4000 B.C. and because the higher-order analyses and interpretations depend for their organization upon the early dating of cultivated plants, the final results are inevitably thrown into question. (Dumond, 1987:195)

Dumond thus focused on the critical and central importance of the Guilá Naquitz domesticates—that the entire higher level structure of interpretation and explanation rested squarely on the presence of squash and gourd in the cave's preceramic zones. He then suggested that they might, in fact, be more recent in age. This potentially devastating possibility, that the Guilá Naquitz domesticates might not be nearly as old as initially thought, loomed even larger several years later, when direct accelerator mass spectrometer (AMS) radiocarbon dating of the earliest known archaeological maize (*Zea mays*), from Coxcatlán and San Marcos Caves in Tehuacán, showed it to be not 7,000 years old, as long supposed, but dated to only 4700 B.P. (3500 calibrated [cal.] B.C.) (Long et al., 1989). More recently, Lawrence Kaplan's program of direct AMS radiocarbon dating of early domesticated *Phaseolus* beans from throughout the Americas also resulted in much more recent age determinations (Kaplan 1993, 1995). In addition, direct AMS radiocarbon dates on early bottle gourd and domesticated *Cucurbita* specimens from Romero's and Valenzuela's caves in Ocampo, Tamaulipas, indicated that they first appeared in the Ocampo archaeobotanical sequence, not at 9000 B.P. as originally claimed, but by about 4500–4300 cal. B.C. (Smith, 1997b).

Taken together, these AMS radiocarbon dating projects served to focus increasing interest on Guilá Naquitz and to reinforce the possibility raised by Dumond—that the domesticated squash recovered from the preceramic layers of this small and shallow Oaxacan cave were not as old as initially thought. Of the five caves that had yielded all the evidence for initial plant domestication in Mesoamerica, after all, four had been restudied (Coxcatlán, Romero's, San Marcos, Valenzuela's caves), and in each of these four caves the earliest domesticates had turned out to be much younger than originally claimed. The single cave remaining to be restudied, and the only one with early but as yet not directly dated domesticates, was Guilá Naquitz. It is not surprising that as a result several scholars began to explicitly question the age of the Guilá Naquitz squash and, with it, the overall antiquity of the initial transition to food production in Mesoamerica.

Citing Dumond, for example, along with several of the AMS studies mentioned above, Fritz (1994:306) concluded that the bottle gourd and domesticated *Cucurbita pepo* at Guilá Naquitz could well be intrusive into the preceramic zones from Zone A (Late Classic): "Even if Zone B is as old as Flannery suggests, Postclassic activities in combination with nonhuman animal vectors could be responsible for all the cultigens found in Preceramic levels at this small, shallow site." In addition to calling for the direct AMS radiocarbon dating of any and all proposed early domesticate specimens, including those from Guilá Naquitz, Fritz also underscored the necessity of clearly establishing the domesticated versus wild status of archaeobotanical domesticate candidates (Fritz, 1994:308) and argued that, once this new rule of evidence (direct dating of clearly domesticated specimens) was employed, the transition to food production in Mexico and "the first American Farmers" would be shown to be much younger than was believed to be the case. I also followed this general line of reasoning in several publications (Smith, 1995a, 1995b), outlining the need for a more exacting standard of evidence for plant domestication in the Americas, one that centered on the direct dating of specimens exhibiting clear morphological changes associated with domestication, while also proposing that the initial domestication of plants in Mesoamerica likely did not take place before about 6,500 years ago. As part of this proposed shortened chronology for plant domestication in Mesoamerica, the twin issues of the true antiquity of the Guilá Naquitz squash and its domesticated versus wild status were raised, and I suggested that it seemed "quite possible that, with the exception of the potentially intrusive zone B1 squash specimens from Guilá Naquitz, the earlier Zone C–D material is all from a species of wild gourd" (Smith, 1995b:179). This question of the domesticated versus wild status of the *Cucurbita* specimens from Guilá Naquitz remained open because, in the initial analysis of the material (Whitaker and Cutler, 1986), there was no mention of the morphological criteria that had been used to distinguish the seeds, peduncles, and rind fragments of domesticated *C. pepo* squash from the numerous remains of a wild *Cucurbita* gourd that were scattered throughout the preceramic zones of the cave.

These publications (Fritz 1994; Smith 1995a, 1995b), along with the earlier critique by Dumond, made it clear that, even a full decade after the publication of the Guilá Naquitz volume, three key questions still remained to be resolved: (1) What was the degree of stratigraphic integrity of the preceramic zones in Guilá Naquitz cave? (2) Which specimens in the Guilá Naquitz preceramic *Cucurbita* assemblage could be confidently and unequivocally identified as representing domesticated *C. pepo* squash? (3) What was the true antiquity of the specimens of any domesticated *Cucurbita pepo* that could be identified in the preceramic-period archaeobotanical assemblage of Guilá Naquitz cave?

In this chapter I address and answer each of these three questions, beginning with the complicated issue of the degree of disturbance of the Archaic-period occupations of Guilá Naquitz cave.

THE STRATIGRAPHIC INTEGRITY OF GUILÁ NAQUITZ

In raising questions about the general stratigraphic integrity of the Archaic-period levels in Guilá Naquitz, Dumond (1987) focused on the likelihood of displacement of older Zone D material, including charcoal that yielded carbon 14 (^{14}C) samples, up into Zones B1–B3, and both Fritz (1994) and Smith (1995b) suggested the possible downward displacement of material, including domesticates, into Archaic levels from overlying Zone A contexts. As briefly discussed above, the suspicions that materials had been displaced vertically between zones were based on a number of factors. Among the most important were the shallow nature of the cave's deposits, along with the potential for both human and rodent disturbance, and the related problems with dating by association (recently underscored by the results of direct AMS dating of early domesticates in other caves). These suspicions about vertical displacement also seemed to be supported by the radiocarbon dates that had been obtained on charcoal samples from the Archaic-period zones of Guilá Naquitz (Table 1). Flannery (1986:175) had considered only 6 of

Table 1. Radiocarbon Samples from Guilá Naquitz Cave Reported by Flannery (1986:Table 14.1)

Zone	Square	Lab and number[a]	Date[b] (years B.C.)	Nature of sample	Comments
B2	D8	SI-515	6670 ± 160	Fragments of charcoal, not identified	Acceptable
B2+3	E6	GX-0784	6910 ± 180	Fragments of charcoal, not identified	Acceptable
C	F6	GX-0873	7280 ± 120	Wood charcoal, not identified	Acceptable
	D10	M-2097	7450 ± 300	Fragments of charcoal, legume tree	Acceptable
	F9	M-2100	4030 ± 220	Fragments of charcoal, too small to identify	Too young
D	F5	GX-0783	7840 ± 240	Wood charcoal, not identified	Acceptable
	E5	M-2099	8750 ± 350	Fragments of charcoal, not identified	Acceptable
	D10	M-2098	2350 ± 180	Charcoal, *Pinus* sp.	Too young
E	D10	GX-0872	5545 ± 90	Fragments of charcoal, too small to identify	Too young
	F8	M-2101	4350 ± 220	Fragments of charcoal, *Quercus* and *Acacia*	Too young

[a] Samples were submitted to three laboratories: Geochron (GX), the Smithsonian Institution (SI), and the University of Michigan (M).

[b] All dates are calculated from 1950, using the 5570 half-life.

these 10 dates to be acceptable. The 4 dates labeled as "too young" (Table 1) suggested the possibility of substantial vertical movement of materials between zones, despite the apparent clear and consistent separation of strata that was observed during excavation (Flannery, 1986:65–95).

In the past several years, 27 additional radiocarbon age determinations have been obtained by Kent Flannery, Lawrence Kaplan, and the author for carbonized and uncarbonized plant remains from the Archaic-period zones of Guilá Naquitz (Table 2). These additional dates provide important chronological and stratigraphic context and clarification of the 10 original Guilá Naquitz dates, particularly the 4 that seemed to be "too young" and out of sequence. When all 37 of the radiocarbon dates now available from the Archaic-period zones (B1–E) of Guilá Naquitz are calibrated and placed in ascending order, irrespective of their stratigraphic zone assignments (Table 2), a number of interesting patterns emerge.

Table 2. Guilá Naquitz Radiocarbon Dates in Ascending Order of Dendrocalibrated Age

Labratory sample number	Zone	Sq	Conventional ^{14}C age in radiocarbon years: Years B.P.	Conventional ^{14}C age in radiocarbon years: Years B.C.	Dendrocalibated age in calendar time scale (B.C.): Intercept[a]	Dendrocalibated age in calendar time scale (B.C.): 2-sigma age range[b]	Material	Comments
B97236	B1	D5	2530 ± 60	580	775	810–415	Bottle gourd rind	GN 15, AMS, INAH 31
M2098	D	D10	4300 ± 180	2350	2900	3485–3465 3375–2460	Charcoal (*Pinus* sp.)	
Middle Archaic Occupation Cluster								
M2100	C	F9	5980 ± 220	4030	4855	5330–4360	Charcoal fragments	
B76873	E	F7	6160 ± 70	4180	5065	5255–4920	Wood charcoal	GN 14
M2101	E	F8	6300 ± 220	4350	5255	5600–4750	*Quercus* and *Acacia* Charcoal fragments	
B76872	E	H9	6570 ± 70	4620	5455	5590–5335	Wood charcoal	GN1
Zones B1–B3 Early Archaic Occupation Cluster								
B97240	B(B1)	B9	6980 ± 50	5030	5805	5950–5705	Squash peduncle scar	GN 20, AMS, INAH 58
B97238	B1	C11	7280 ± 60	5330	6115	6195–5980	Squash peduncle scar	GN 18, AMS, INAH 42
B74751	B(B1)	D7	7310 ± 70	5360	6145	6225–5985	Wood charcoal	GN 8
B97239	B(B1)	B9	7340 ± 60	5390	6165	6235–6015	Squash peduncle	GN 19, AMS, INAH 58
B74748	B(B1)	D7	7450 ± 60	5500	6225	6400–6160	Wood charcoal	GN 1, AMS
GX872	E	D10	7495 ± 90	5540	6360	6460–6150	Charcoal fragments	
AA13336	B1	C11	7506 ± 90	5560	6370	6400–6220	Wild *Phaseolus*	AMS
AA15007	B1	C11	7583 ± 62	5633	6400	6480–6220	Wild *Phaseolus*	AMS
B91406	B1	C11	7610 ± 60	5660	6425	6485–6355	Squash seed	GN- D, AMS, INAH 7
B91405	B1	C11	7690 ± 50	5740	6465	6570–6415	Squash seed	GN- B, AMS, INAH 7
B100763	B2	C11	7710 ± 50	5760	6475	6585–6425	Squash seed	GN 21, AMS, INAH 5
B91404	B1	C11	7720 ± 60	5770	6480	6610–6420	Squash seed	GN- A, AMS, INAH 7

B76868	C	E7	7920 ± 80	5970	6705	7025–6535	Wood charcoal	GN 9
B97237	B2	E10	7940 ± 60	5990	6735	7020–6595	Bottle gourd rind	GN 16, AMS, INAH 45
SI515	B2	D8	8620 ± 160	6670	7575	7995–7325	Charcoal fragments	
Zones C–D Early Archaie Occupation Cluster								
GX784	B2+3	E6	8860 ± 180	6910	7945	8235–7565	Charcoal fragments	
B76869	D	E7	8870 ± 80	6920	7955	8045–7695	Wood charcoal	GN 10
B100762	C	F9	8900 ± 50	6950	7970	8030–7915	Bottle gourd rind	GN 17, AMS, INAH 63
B100764	C	E9	8910 ± 50	6960	7975	8035–7920	Squash seed	GN 22, AMS, INAH 14
B100766	B1	E11	8990 ± 60	7040	8025	8085–7955	Squash seed	GN 24 , AMS, INAH 23
B74750	D	E7	9010 ± 60	7060	8035	8095–7965	Wood charcoal	GN 3, AMS
GX873	C	F6	9230 ± 120	7280	8215	8510–8025	Wood charcoal	
B76870	D	F8	9350 ± 80	7410	8395	8570–8140	Wood charcoal	GN 11
B76871	D	F8	9380 ± 70	7430	8410	8580–8230	Wood charcoal	GN 12
M2097	C	D10	9400 ± 300	7450	8425	9150–7915	Legume charcoal frags.	
B74626	D	F8	9520 ± 60	7570	8580	8955–8425	Wood charcoal	GN 5, AMS
B74627	E	F7	9580 ± 60	7630	8640	8995–8495	Wood charcoal	GN 6, AMS
B74625	D	F8	9580 ± 70	7630	8640	9005–8465	Wood charcoal	GN 4, AMS
GX783	D	F5	9790 ± 240	7840	9035	10,365–8365	Wood charcoal	
B74749	C	E7	10340 ± 60	8390	10,355 10,270 10,210	10,835–9940	Wood charcoal	GN 2, AMS
M2099	D	E5	10700 ± 350	8750	10,890	11795–9150	Charcoal fragments	
B74628	E	G4	14420 ± 60	12,470	15,320	15,705–14,945	*Taxodium* wood uncarbonized	GN 7, AMS

[a]Intercept between the conventional ^{14}C age and the dendrocalibrated calendar time scale, in calendar years B.C., determined using the Pretoria calibration procedure program of Beta Analytic.

[b]Two-sigma dendrocalibrated age range for specimens, in calendar years B.C.

Significantly, none of the 37 dated archaeobotanical specimens from Zones B1–E turned out to be Late Classic in age (A.D. 620–740). Contrary to the suggestions of both Fritz and Smith, there is no evidence for any unrecognized infiltration of Zone A materials down into the Archaic-period levels of the cave (see Flannery, 1986:93, for a discussion of the efforts made during excavation to recognize such disturbed deposits).

During the excavation of Guilá Naquitz, isolated small scatters of ceramic sherds were recorded at the boundary of Zones A and B1, indicating occasional and ephemeral visits to the cave during the Formative period (Flannery, 1986:8,95). Further evidence of this Formative-period presence was provided by a bottle gourd rind fragment recovered from an excavation unit along the back wall of the cave, which yielded a Monte Albán I date of 775 cal. B.C. (Table 2). With the exception of this single Formative-period date, all the other 36 age determinations for specimens from Zones B1–E are associated with the Archaic-period occupations of the cave.

This sequence of 36 preceramic ^{14}C dates also provides further evidence of brief Middle Archaic (c. 5500–4900 cal. B.C.) occupations of Guilá Naquitz hinted at by three of the four original dates initially labeled as "too young" (Table 1). There are now a total of five Middle Archaic dates for the cave, with four of these forming a temporal and spatial cluster, ranging from 5455 to 4855 cal. B.C. in age and occurring in four 1-by-1-m excavation units. Three of these are contiguous (Figure 3, Table 2). This Middle Archaic spatial and temporal cluster in turn suggests a possible straightforward solution to the problem posed by the seemingly out-of-sequence "too young" dates from Guilá Naquitz. The cluster is situated on the margin of most of the evidence for earlier Archaic-period occupation and activity areas (see Flannery, 1986:Figs. 28.10, 28.17, 28.24, 28.31) and might represent ephemeral and previously unrecognized lenses of Middle Archaic occupation. Although there was apparently some vertical compression and commingling of materials from these Middle Archaic occupations with debris from earlier underlying habitation layers in this portion of the cave, such was not the case in other parts of Guilá Naquitz, judging from two other spatial-temporal clusters of ^{14}C dates.

The first of these is the "Zones B1–B3" cluster of 12 dates ranging in age from 6480 to 5805 cal. B.C., from four excavation units toward the back of the cave (Table 2, Figure 4). Eleven of the dates in this cluster come from Zones B1–B3 contexts, indicating a high level of stratigraphic integrity for the sequence of occupational episodes (c. 6500–5800 cal. B.C.) that centered on activity areas I–III and the associated hearth (Feature 14) in this part of the cave (Flannery, 1986:Fig. 28.31). On the basis of the nonoverlap of various dates at the 2-sigma (95% probability) range, this Zones B1–B3 temporal and spatial cluster could represent four separate occupational episodes.

Predating this Zones B1–B3 cluster of dates by about 1,500 years, a second

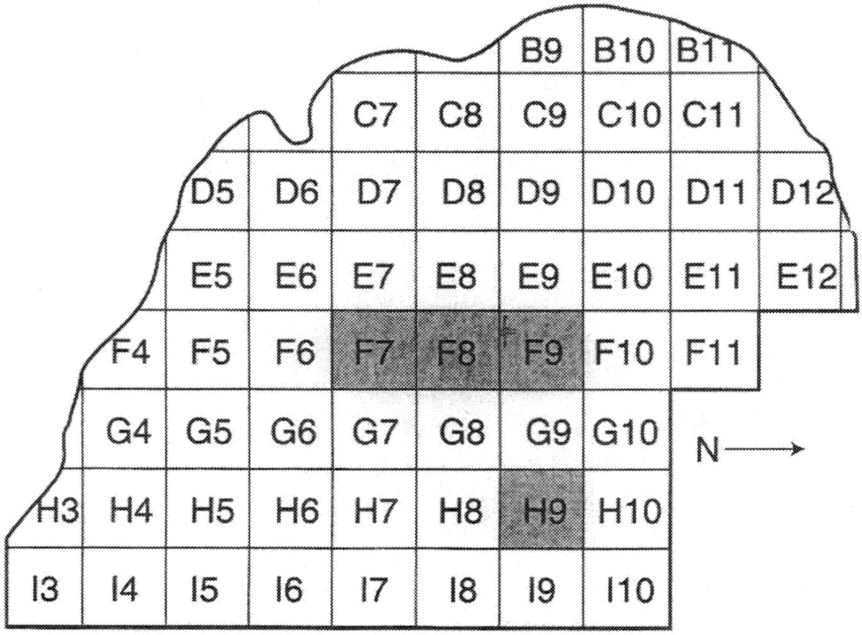

Figure 3. The four 1-x-1-m excavation units in Guilá Naquitz cave that yielded the Middle Archaic spatial-temporal cluster of four dates (dendrocalibrated calendar time scale intercepts in years B.C.: 5455, 5255, 5065, 4855; see Table 2).

series of 13 dates, ranging in age from 8640 to 7945 cal. B.C., is spread across a total of nine contiguous excavation units (Table 2, Figure 5) that also contained evidence of Zones C and D activity areas (Flannery, 1986:Figs. 28.17, 28.24). Ten of these thirteen dates were from Zones C and D contexts, one was from a Zone E context, and two were from Zones B1–B3 contexts. Several important points can be made about this Zones C–D series of dates. First, and most important, it is possible, working within the 2-sigma age range given for each date, to arrange them in stratigraphic sequence: The oldest possible value for the Zone E date in this cluster (8995 cal. B.C.) is older than the youngest possible age for the oldest Zone D date (8505 cal. B.C.), and the oldest possible value for the youngest Zone D date (8045 cal. B.C.) is, in turn, older than the youngest possible value for the oldest Zone C date (7556 cal. B.C.). Thus although the dates in this cluster do suggest the possibility of some mixing of materials from Zones C and D occupations, they at the same time do not preclude the possibility of total stratigraphic integrity. In light of the overlap at the 2-sigma range, this series of dates may represent two occupational episodes.

Three dates (7575, 6735, 6705 cal. B.C.) fall into the 1,500-year-long gap (c.

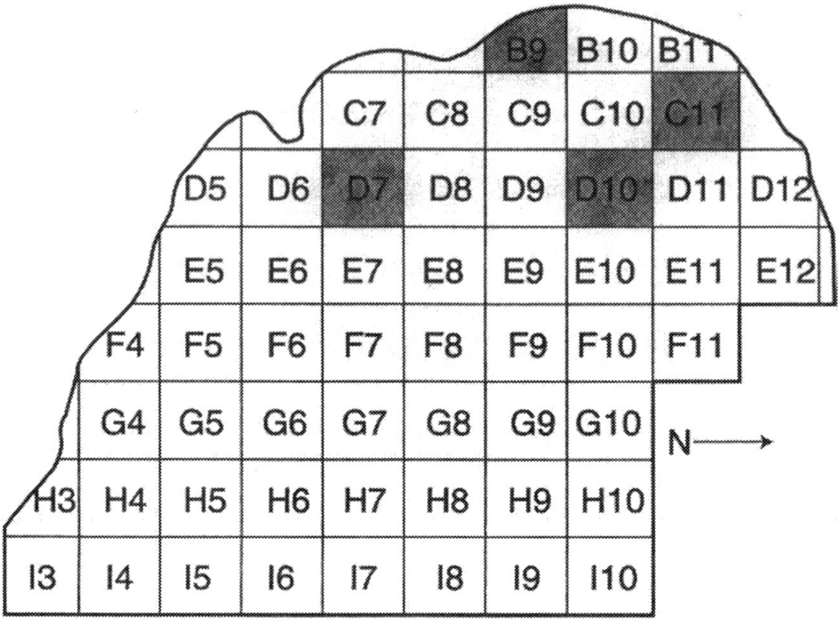

Figure 4. The four 1-x-1-m excavation units in Guilá Naquitz cave that yielded the Zone B Early Archaic spatial-temporal cluster of 12 dates (dendrocalibrated calendar time scale intercepts in years B.C.: 6480, 6475, 6465, 6425, 6400, 6370, 6360, 6225, 6165, 6145, 6115, 5805; see Table 2).

8000–6500 cal. B.C.) between the Zones B1–B3 and Zones C–D date clusters (Table 2). Given the 2-sigma age ranges for these three dates, they could be either outliers for the clusters on either side or, alternatively, could actually indicate occasional human presence in the cave between 8000–6500 cal. B.C. Finally, a series of four dates from four squares along the southern wall of the cave (Table 2) document the presence of humans at Guilá Naquitz before 9000 cal. B.C.

What then can be said in summary of the stratigraphic integrity and sequence of Archaic-period occupation of Guilá Naquitz, based on the 37 radiocarbon dates now available for the cave? In my opinion, these dates strongly suggest that Guilá Naquitz was inhabited primarily during two seven-century spans, from c. 8700–8000 cal. B.C. (Zones C-D), and then again from c. 6500–5800 cal. B.C. (Zones B1–B3). During each of these 700-year-long periods, small family groups occupied Guilá Naquitz in the fall of the year on an unknown number of occasions. The available dates also indicate a more sporadic and ephemeral pattern of Archaic-period occupation of the cave, both earlier in time, before to 9000 cal. B.C., and later, after 5800 cal. B.C. Finally, and most important, I argue that this large set of radiocarbon age determinations also indicates that the record of habitation of

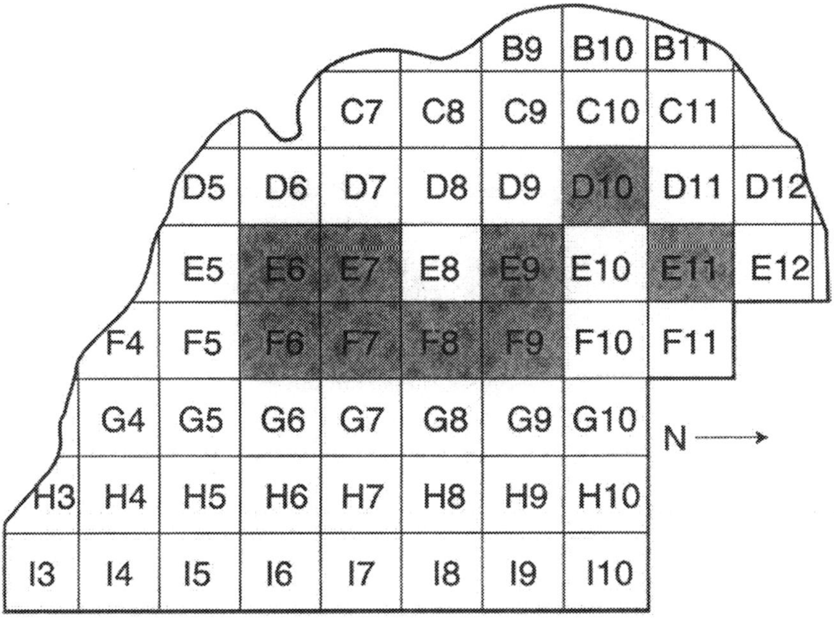

Figure 5. The nine 1-x-1-m excavation units in Guilá Naquitz cave that yielded the Zones C–D Early Archaic spatial-temporal cluster of 13 dates (dendrocalibrated calendar time scale intercepts in years B.C.: 8640, 8640, 8580, 8425, 8410, 8395, 8215, 8035, 8025, 7975, 7970, 7955, 7945; see Table 2).

Guilá Naquitz during the two major periods of occupation was substantially intact when excavated, and there was very little unrecognized vertical displacement of materials between the major zones of occupation.

THE GUILÁ NAQUITZ CUCURBIT ASSEMBLAGE

The cucurbit materials from Guilá Naquitz were originally studied by Thomas Whitaker and Hugh Cutler (Whitaker and Cutler, 1971, 1986), widely respected botanists and world authorities on squash and the analysis of *Cucurbita* materials from archaeological contexts. The reanalysis of the Guilá Naquitz cucurbit collection presented here can in many respects be considered an update and expansion of their earlier research, while also occasionally taking issue with their analysis and results. Whitaker and Cutler did not have the benefit of being able to directly determine the age of the cucurbit material they were analyzing through accelerator mass spectrometer (AMS) small sample ^{14}C dating; and clear standards for establishing the domesticated status of archaeological squash specimens had not yet been fully formulated. This study addresses these key issues. Twelve AMS ^{14}C age determinations for cucurbit specimens from Guilá Naquitz are reported, and a detailed description and analysis of the cave's cucurbit assemblage are presented, with particular attention paid to recognizing the presence of potential morphological markers of domestication. At the same time, information is presented on a variety of wild species of *Cucurbita* gourds, both ancient and modern, providing a baseline of comparison for the Guilá Naquitz specimens. In addition to focusing on these two basic standards of evidence for establishing the age and domesticated status of a crop plant (direct AMS age determination and clearly identified morphological changes linked to the adaptive syndrome of domestication), an effort is also made, following Newsom et al. (1993:87–89) and Cowan and Smith (1993:35–51), to consider and compare multiple lines of evidence for morphological changes in reproductive propagules associated with domestication, including changes in rind thickness, seed size, peduncle (fruit stem) size, and fruit size, shape, and color.

Bottle Gourd *(Lagenaria siceraria)* Rind Fragments

In addition to the larger assemblage of material assignable to the genus *Cucurbita*, as well as the four seeds of the wild coyote melon (*Apondanthera*) recovered during excavation of the preceramic zones of Guilá Naquitz, a small number of rind fragments of another cucurbit—bottle gourd (*Lagenaria siceraria*)—also were identified. In their initial analysis, Whitaker and Cutler listed a total of seven bottle gourd rind fragments, three from Zone C contexts and four from Zones B1–B3 (Whitaker and Cutler, 1971:125, 1986:278–279).

In the present study, however, only three bottle gourd rind fragments were identified in the Guilá Naquitz cucurbit collection, currently curated in the Laboratorio de Paleobotánica, Instituto Nacional de Antropología e Historia (INAH), Mexico City. All three of these rind fragments were uncarbonized; all were distinctive dark reddish brown to brown in surface color; and all exhibited the distinctive cross-section pattern of elongated, loosely organized inner cells diagnostic for *Lagenaria siceraria* (Figure 6) (Cutler and Whitaker, 1961:479; Roberts, 1995:7–8; Smith 1997b). Direct AMS ^{14}C dates obtained for these three bottle gourd rind fragments indicate that, whereas one belongs to a Formative-period (Monte Albán I) occupation of the cave, the other two fall comfortably into the preceramic occupation (Table 3). The earliest of these dates (7970 cal. B.C.) confirms previous age estimates for Guilá Naquitz *Lagenaria* (Flannery, 1986; Pickersgill and Heiser, 1977) and represents the oldest direct date so far obtained for an archaeological specimen of this species. The oldest AMS ^{14}C date for bottle gourd in North America north of Mexico is 6125 cal. B.C., reported for a small and partially intact bottle gourd recovered from a burial context at the Windover site

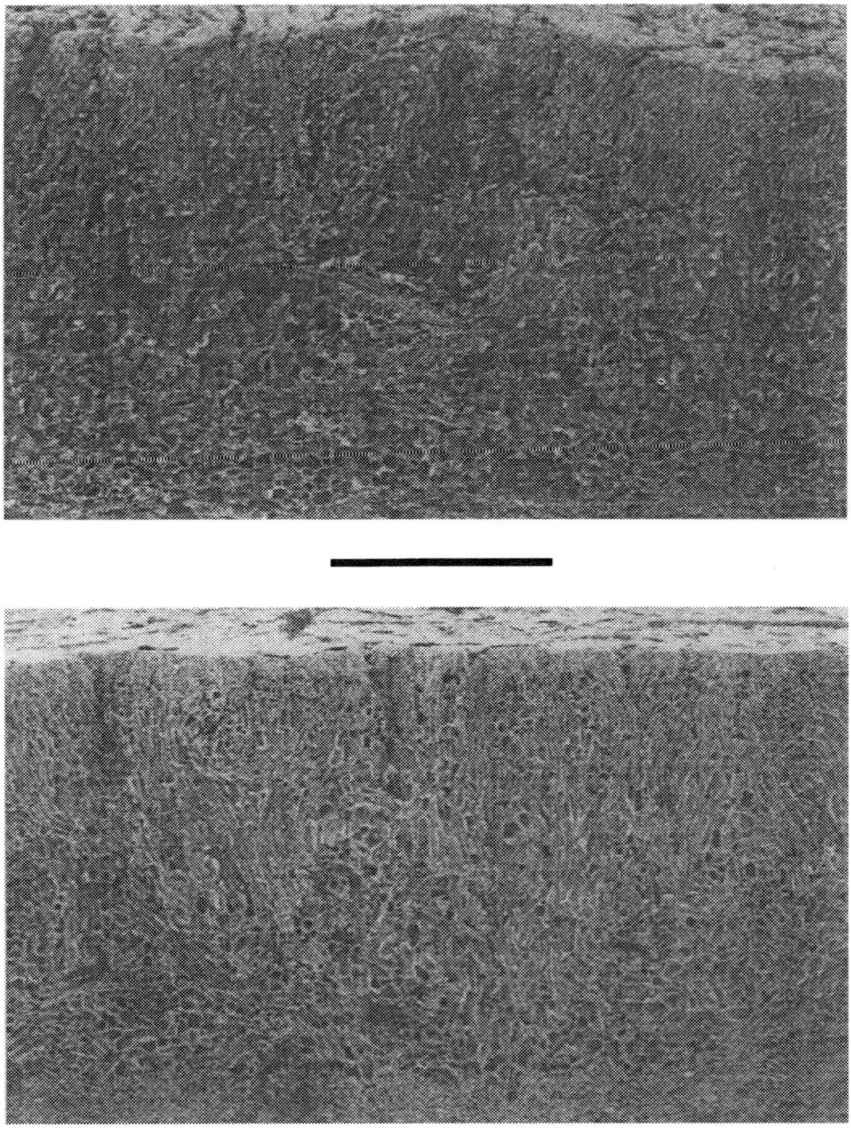

Figure 6. Scanning electron micrographs of cross-section of bottle-gourd rind fragments from Guilá Naquitz, showing the distinctive pattern of elongated and loosely organized inner cells diagnostic for *Lagenaria siceraria.* (above) Bottle gourd rind fragment from Square D5, Zone B1 (INAH 31), which was evidently intrusive into the preceramic habitation layers of the cave, based on an AMS date of 775 cal. B.C. (below) Bottle gourd rind fragment from Square F9, Zone C, which yielded an AMS date of 7970 cal. B.C. (Tables 2, 3). Scale bar = 1 mm.

Table 3. Bottle Gourd Rind Fragments from Preceramic Zones of Guilá Naquitz.

INAH catalog number	Zone and square	Rind thickness (mm)	Rind weight (gm)	AMS age determination (years B.C.)[a]
31	B1 / D5	2.2	.21	775
45	B2 / E10	4.6	.03	6735
63	C / F9	2.2	.65	7970

[a]Dates shown represent dendrocalibrated age intercept in calendar time scale (years B.C.). See Table 2 for more information.

in eastern coastal Florida (Doran and Newsom, 1990). The Windover and Guilá Naquitz direct AMS dates, along with other dates by association, document the early Holocene human use of bottle gourd in the Americas, but do not answer the long-unresolved questions about whether these early bottle gourds represent wild or domesticated plants or something in between (Flannery, 1973:303, 306; Heiser, 1990; Pickersgill and Heiser, 1977).

Any attempt to make a case for domesticated bottle gourd in an early archaeological context on the basis of morphological criteria (e.g., increased seed size) has so far been hampered by both the absence of bottle gourd seeds in early temporal contexts and the absence of a wild baseline of comparison. Because no historically documented or present-day free-living *Lagenaria* populations in either Africa (its possible homeland?) or on this side of the Atlantic have yet been identified as being "truly wild" (Heiser, 1990:113), the size and shape characteristics and range of variation for wild bottle gourd seeds, stems, and fruits remain undocumented, making it impossible to establish whether archaeological specimens are domesticated or wild (Heiser, 1990:113). In the absence of any clear morphological criteria for establishing the wild versus domesticated status of bottle gourd in early archaeological contexts, an implicit geographical-range argument often seems to be employed to assign domesticated status by default: Because wild bottle gourds do not grow in the region today, the archaeological site in question is apparently outside the (perhaps nonexistent) present-day natural range of the wild progenitor, and any *Lagenaria siceraria* found at the site must have been introduced and husbanded, hence in some sense "domesticated." In spite of the "default domesticate" solution sometimes offered for early Holocene bottle gourd in the Americas, an adequate understanding of its early history as a wild plant and its initial domestication will only come only through documentation of modern wild populations, and through genetic analysis of *Lagenaria* worldwide.

Cucurbita Rind Fragments

A total of 276 rind fragments in the Guilá Naquitz collection curated by INAH were assigned to the genus *Cucurbita*, based on their light tan to brown surface color, thinness (< 2 mm), and characteristic cross-section cell structure—isodiametric with regular configuration (Roberts, 1995:7). Only a few of the *Cucurbita* rind fragments showed any evidence of burning; they were uniformly smooth and unlobed. Whitaker and Cutler, however, recorded only 55 *Cucurbita* rind fragments in their initial analysis of the preceramic cucurbit assemblage from Guilá Naquitz (C. Smith, 1986:266; Whitaker and Cutler, 1971:125, 1986:278). To an unknown extent, this difference in rind count may be a result of breakage that occurred subsequent to the initial analysis.

When compared in terms of maximum thickness measurements, the *Cucurbita* rind-fragment populations from preceramic Zones D, C, and B1–B3 of the cave show an interesting pattern (Figure 7). Although the Zone D (n = 30) and C (n = 168) populations are similar to each other in terms of mean thickness (.84 mm, .81 mm), as well as in range (0.5–1.6 mm, 0.3–1.7 mm) and standard deviation (.22, .22) values, they differ substantially from the Zones B1–B3 population, which has larger mean (1.15 mm), range (0.5–2.0 mm), and standard deviation (.31) values (Figure 7). When combined with the bimodal shape of the Zones B1–B3 rind-thickness distribution curve, these larger values suggest that a second, thicker-rind form of *Cucurbita* has joined a thinner-rind form. This appearance of a thicker-rind *Cucurbita* could be the result of deliberate human selection for thicker walled fruit and could indicate the presence of a domesticated squash in Zones B1–B3 of the cave. The Zones B1–B3 *Cucurbita* rind assemblage, however, is not that different in terms of thickness values from modern species of wild *Cucurbita* gourd (Figure 7) and falls far short of the 2-mm minimum value for rind thickness proposed for clearly domesticated *Cucurbita* (Cowan and Smith, 1993:40; Newsom et al,, 1993:91). As a result, this increase in rind thickness is at most suggestive only of the possible presence of a thicker-walled domesticated *Cucurbita* squash in Zones B1–B3 of Guilá Naquitz. It does, however, also suggest that perhaps other changes in fruit morphology resulting from deliberate human selection may have accompanied such an increase in rind thickness, and that such changes might be seen in a comparison of the *Cucurbita* peduncles (fruit stems) and peduncle scars (where the peduncle attaches to the fruit) from Zones D, C, and B1–B3 of Guilá Naquitz.

Cucurbita Peduncles and Peduncle Scars

A total of six *Cucurbita* peduncles and partial peduncles from Zones D, C, and B1–B3 of Guilá Naquitz were present in the INAH collection. Although none of these peduncles was still attached to the fruit, and none had convex discs of rind still attached to the peduncle base, as was observed for specimens from the Ocampo Caves (Smith, 1997b), two were found to fit fruit-rind fragments that carried corresponding stem attachment scars. These two stem/fruit pairs were probably still attached when excavated (Figure 8). In addition to the six detached peduncles and two peduncle fruit-end fragment pairs, another eight unmatched fruit fragments

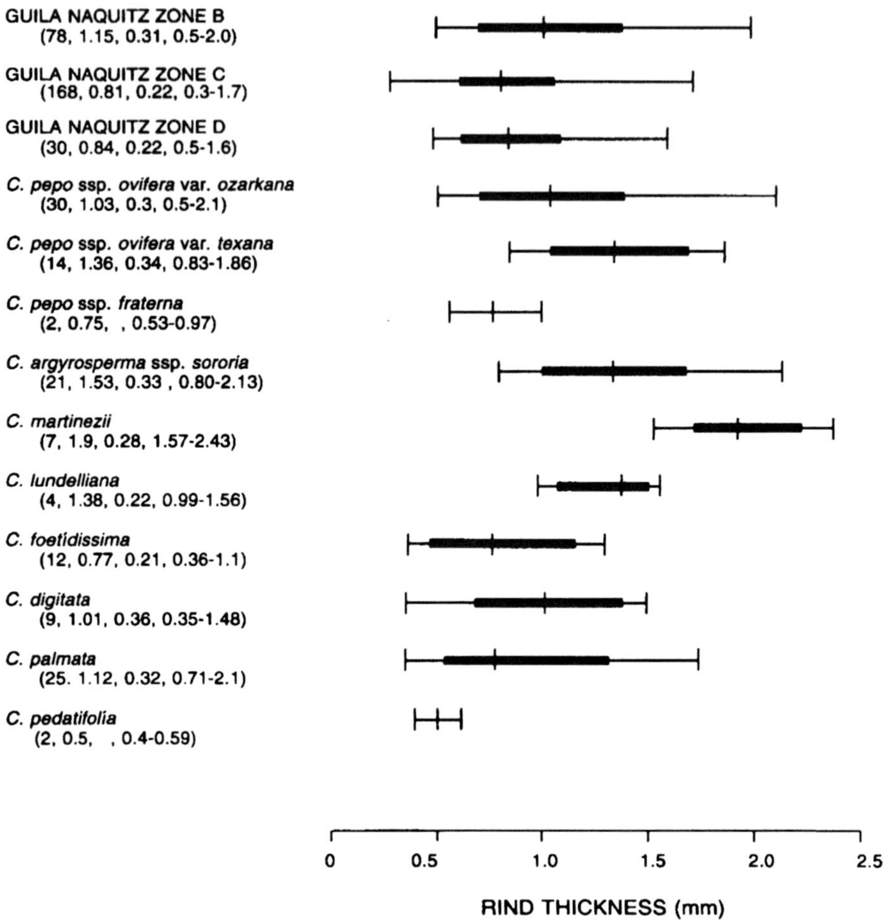

Figure 7. Thickness measurements for *Cucurbita* rind fragments recovered from preceramic habitation zones of Guilá Naquitz, compared with thickness values for six taxa of modern wild *Cucurbita* gourds. Based on rind measurements obtained on cataloged specimens from the following museum and herbarium collections—Field Museum of Natural History: 108583, 1109554, 1262357, 1883666; Illinois State Museum: 671; Missouri Botanical Garden: 406218, 1964873, 2323729, 2393083, 2467558, 2483791, 2587700, 3015602, 3201096, 3234773, 3241789, 3264716, 3379643, 3379647, 3379648, 3379650; National Museum of Natural History: 553 (Fruit), 29350, 453271, 64648, 1419024, 1468353, 1587372, 1747269, 1768692, 2022009; Research collections of D. Decker-Walters, B. Smith. Numbers in parentheses are sample size, mean, standard deviation, and range.

having peduncle scars were also present in the collection, along with five fruit fragments having blossom scars (Table 4). With a few differences, this peduncle and fruit end fragment count compares closely to that of Whitaker and Cutler (1986:278–279).

Figure 8. *Cucurbita* peduncle and matching fruit-end fragment with peduncle scar from Guilá Naquitz cave (INAH 2), which exhibit morphological characteristics typical of wild morphotype *Cucurbita* gourds: small size (maximum basal peduncle diameter 11.4 mm), smooth, somewhat depressed circular peduncle scar suggesting oblate fruit form, and absence of basal lobing of peduncle or lobing of fruit.

In terms of both size and general morphology, 8 of the 16 specimens in this small assemblage of *Cucurbita* fruit stems and stem scars are comparable to present-day taxa of wild *Cucurbita* gourds. Although these specimens from Guilá Naquitz perhaps represent the wild basal progenitor of the Mexican lineage of *C. pepo* squash (*C. pepo* ssp. *pepo*) (Decker, 1986), it is difficult to attempt any taxonomic assignment beyond the genus level because modern wild *Cucurbita* gourds exhibit considerable similarity in form. Wilson (1989) has documented the general morphological similarity in reproductive propagules of a range of different free-living *Cucurbita* species in Mexico, which he characterized as "relatively uniform and phenetically unified by a suite of morphological characteristics" (Wilson, 1989:618). As a result, because the 10 or so wild *Cucurbita* species so far described (Robinson and Decker-Walters, 1997:71) all produce generally similar hard-walled small fruits having small stems and small seeds, it is as yet difficult to identify archaeological specimens of wild *Cucurbita* peduncle and peduncle scars as to species with any degree of confidence. Although making species assignments difficult, this considerable degree of similarity in terms of fruit and stem size and shape among free-living species of *Cucurbita* does, at the same time, allow for a general wild *Cucurbita* morphotype profile to be developed. This profile, in turn, can be used as a baseline for distinguishing wild from domesticated specimens in archaeobotanical assemblages.

Consider, for example, the measurements of maximum basal peduncle diameter for 11 taxa of wild *Cucurbita* gourds presented in Figure 9. When combined with wild *Cucurbita* gourd rind-thickness measurements (Figure 7) and seed size

Table 4. ***Cucurbita*** **Peduncles and Fruit Fragments Having Peduncle Scars from Guilá Naquitz**

INAH catalog number	Square/ Zone	Specimen description	Maximum basal diameter (mm)	Weight (gm)	Age: Dendro-calibrated intercept
58	B9 / B(B1)	*C. pepo* fruit end peduncle scar	23.6	.33	5805 B.C.
14	E9 / C	Withered, pentagonal peduncle	20.8		
42	C11 / B1	*C. pepo* fruit end peduncle scar	20.2	.19	6115 B.C.
58	B9 / B(B1)	*C. pepo* major/minor lobed peduncle	19.0	.36	6165 B.C.
45	E10 / B2	Pentagonal peduncle scar	16.1		
42	C11 / B1	Pentagonal peduncle scar	15.5		6450 B.C.
42	C11 / B1	*C. pepo* lobed peduncle	14.7		6450 B.C.
63	F9 / C	Globular fruit-end peduncle scar	13.0		
	E10 / C?	Peduncle, estimate from photo	12.4		
37	E9 / B2	Globular fruit-end peduncle scar	11.4		
02	G8 / D	Round peduncle paired with fruit end	11.4		
02	G8 / D	Globular fruit-end peduncle scar	11.3		
	E8 / C	Round peduncle scar estimate, photo	11.0		
52	F9 / Lower C	Round peduncle	8.7		
52	F9 / Lower C	Round peduncle paired with fruit end	8.2		
52	F9 / Lower C	Round fruit-end peduncle scar	8.2		
57	E6 / B2-3	Peduncle scar	7.4		
22	D9 / B2	Peduncle scar	7.4		

data, this peduncle diameter information helps define a general wild *Cucurbita* morphotype profile. Such an initial effort to build a general baseline of comparison can be strengthened considerably, of course, by both increasing the sample size of measured specimens for most taxa listed and expanding the range of wild gourds considered, both past (e.g., Newsom et al., 1993) and present.

At the same time, it is hoped that it will not be too long before present-day populations of the wild progenitors of each of the domesticated species of *Cucurbita* are identified, allowing baseline profiles for species-specific wild ancestor gourds to be defined and employed in place of a more general wild *Cucurbita* gourd profile. Although such narrowly drawn wild progenitor profiles are in place for *Cucurbita argyrosperma* (Merrick and Bates, 1989) and for the eastern North American lineage of *C. pepo* (*C. pepo* ssp. *ovifera*) (Cowan and Smith, 1993; Decker-Walters et al., 1993), the wild ancestors of *C. ficifolia, C. moschata,* and the Mexican lineage of *C. pepo* (*C. pepo* ssp. *pepo*) (Decker, 1986) remain to be discovered (Nee, 1990).

In addition to developing a general wild *Cucurbita* gourd morphotype profile based on modern specimens of a range of different species, it is also important to monitor possible morphological variation in wild species of *Cucurbita* over time by including, when possible, wild comparative materials having considerable time

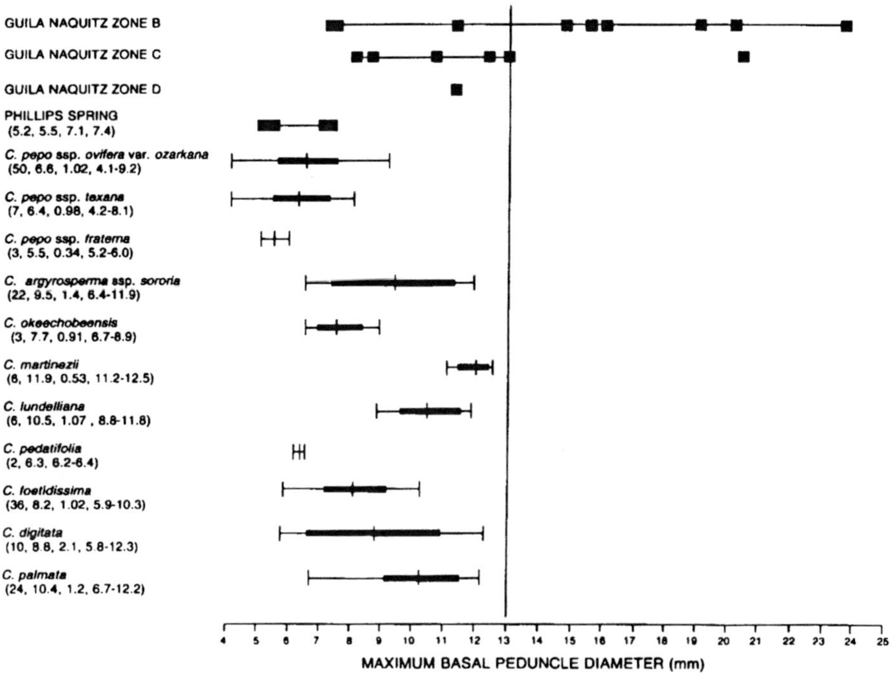

Figure 9. Maximum basal peduncle measurements for *Cucurbita* peduncles from preceramic habitation zones of Guilá Naquitz cave, compared with those of 11 modern taxa of wild *Cucurbita* gourds, along with those from the Phillips Spring site in Missouri. Measurements for Guilá Naquitz specimens are given in Table 4. Statistics given for the 12 reference taxa are sample size, mean, standard deviation, and range. Based on basal peduncle diameter measurements obtained on cataloged specimens from the following museum and herbarium collections—Fairchild Tropical Gardens: DDW-W, 924-0-2; Field Museum of Natural History: 51359, 108583, 1109554, 1262357, 1883666, 1928660, 1972890; Illinois State Museum: 652, 671, 777, 780, 781, 782, 783; Missouri Botanical Garden: 11323, 406218, 1964873, 2323729, 2323761, 2393083, 2467558, 2483791, 2587700, 3015602, 3201096, 3241789, 3264716, 32647, 3379643, 3379647, 3379648, 3379650, 3409542, 3454831, 3534773; National Museum of Natural History: 553 (Fruit), 29350, 64648, 453271, 1419024, 1468355, 1587372, 1747269, 1768692, 2022009.

depth, such as the Florida wet-site collections (Newsom et al., 1993). These assemblages are particularly valuable for comparative purposes when they come from unequivocally noncultural contexts such as the Page-Ladsen site (Newsom et al., 1993), but specimens of probable wild *Cucurbita* gourd from an archaeological sequence are also valuable when they provide a comparative background for recognizing the appearance of morphological changes associated with domestication.

In the Guilá Naquitz *Cucurbita* assemblage, for example, as mentioned earlier, 8 of the 16 peduncles and peduncle scars are comparable in terms of size and morphology with the general wild *Cucurbita* morphotype profile (Table 4, Figure

9). In terms of size, these eight specimens range from 7.4 to 11.4 mm in maximum basal diameter (Table 4), within the range recorded for modern wild *Cucurbita* gourds (4.1–11.6 mm). In addition, a peduncle and peduncle scar illustrated by Whitaker and Cutler (1986:272), but not present in the INAH collections, appear to be compatible with the wild morphotype profile (Table 4, Figure 9). The Guilá Naquitz specimens also fit the wild morphotype profile in a number of other respects. The rind fragments with peduncle scars are smooth with no indication of either lobing or warting, and in all five of the specimens studied the peduncle scar is somewhat depressed, suggesting an oblate fruit form (Figure 8). The peduncle scar illustrated by Whitaker and Cutler (1986:277), however, suggests that a more pronate fruit form also was present. The 10 peduncle scars and peduncles included in this wild morphotype group (including the two specimens illustrated by Whitaker and Cutler) are all roughly circular in outline, with a discernable but not pronounced five-sided angularity (e.g., Figure 8) that is distinct from the prominent pentagonal peduncle scar pattern so characteristic of the domesticated *Cucurbita* species. The peduncles in this group each exhibit a set of five ridges that extend down and angle out into the basal shelf of the peduncle, and basal lobing of these ridges is at most poorly expressed. Smaller intervening and parallel ridges can sometimes be seen between the major ridges, but these do not extend down into the basal peduncular shelf.

In contrast to the group of 10 specimens just described, which conform to the wild *Cucurbita* gourd profile, a second group of seven peduncles and peduncle scars in the Guilá Naquitz assemblage shares a set of morphological characteristics that clearly set it apart from the wild *Cucurbita* morphotype. Ranging in maximum basal diameter from 14.7 to 23.6 mm, the peduncles and peduncle scars in this group are considerably larger than the wild morphotype profile (Table 4, Figure 9), and they all have a distinctive angular pentagonal outline, rather than circular, at the point of peduncle attachment to the fruit.

In addition, four of these seven specimens—two of three peduncles and two of the four fruit-end fragments having peduncle scars—share a distinctive morphology that allows them to be assigned to a particular species of domesticated *Cucurbita*. The distinctive morphology of peduncles of different domesticated squashes has long been recognized as one of the unequivocal ways of establishing the presence of various species of *Cucurbita* (e.g., *C. pepo, C. moschata, C. argyrosperma*) in archaeological assemblages (Cutler and Whitaker, 1961; Smith 1997b). Each of two Guilá Naquitz peduncles, measuring 14.7 and 19.0 mm in maximum basal diameter, has 10 very prominent, ropy, alternating large and small, rounded parallel ridges running perpendicular to the base. In both peduncles, these rounded ridges terminate at the base in lobes that round under before attachment. Although they angle gradually outward, these rounded ridges neither flare abruptly at the base nor coalesce to form an encircling basal peduncular ridge or shelf. Interridge creases extend through to the bottom of the peduncle. This distinctive pattern of 10 deeply grooved, alternating large and small peduncular ridges differs substantially from modern *C. moschata* and *C. argyrosperma*, while closely conforming in terms of peduncle morphology to modern *C. pepo* (Cutler and Whitaker, 1961:476; Smith, 1997b). This similarity can be seen when a Guilá Naquitz peduncle is compared to a peduncle of a modern cultivated type of *C. pepo* ssp. *pepo*, the major lineage of *C. pepo* that originated in Mexico (Figure 10).

This distinctive *C. pepo* pattern of 10 deeply grooved alternating major and minor ridges also carries through onto two of the fruit-end fragments having peduncle scars (Figure 11). Measuring 20.8 and 23.6 mm in maximum diameter, these two peduncle-end fruit fragments have pentagonal peduncle scars, with major peduncle ridge lobes forming the points of the pentangle and the intervening minor ridge lobes located midway along the five sides of the peduncle scar (Figure 11). These fruit-end fragments thus strengthen the case for the presence of a domesticated *Cucurbita* in the Guilá Naquitz preceramic cucurbit assemblage because they document both an increase in peduncle size and a shift to pentagonal

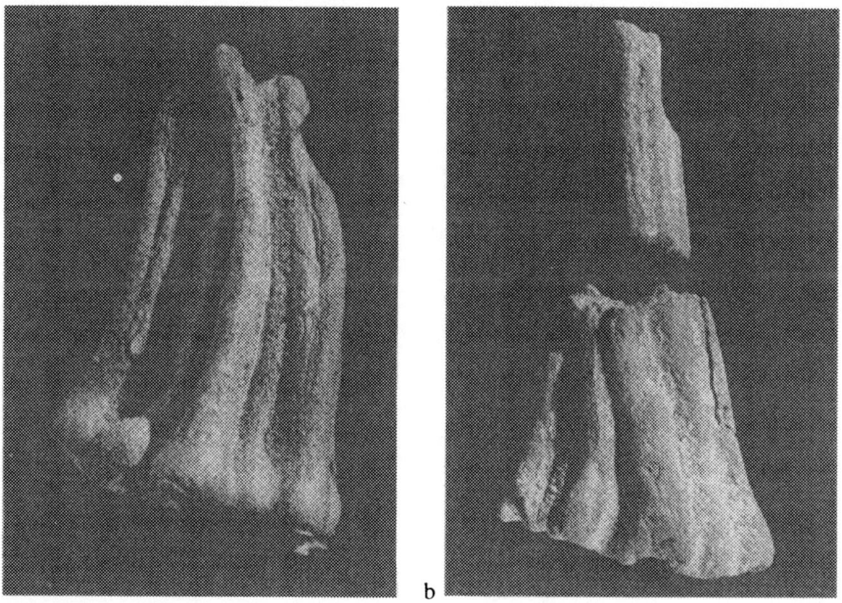

a b

Figure 10. Two peduncles that exhibit morphological characteristics distinctive for domesticated *C. pepo*: prominent, ropy, alternating large and small, rounded parallel ridges that terminate at the base in lobes that round under before attachment, with interridge creases extending through to the bottom of the peduncle. (a) Modern peduncle of *C. pepo* ssp. *pepo* var. "Odessa," 22.3 mm maximum basal diameter, provided by Glen Drowns of Seed Savers exchange. (b) Peduncle from Zone B of Guilá Naquitz (INAH 58), 19.0 mm. maximum basal diameter, 6165 cal. B.C. (Tables 2, 4).

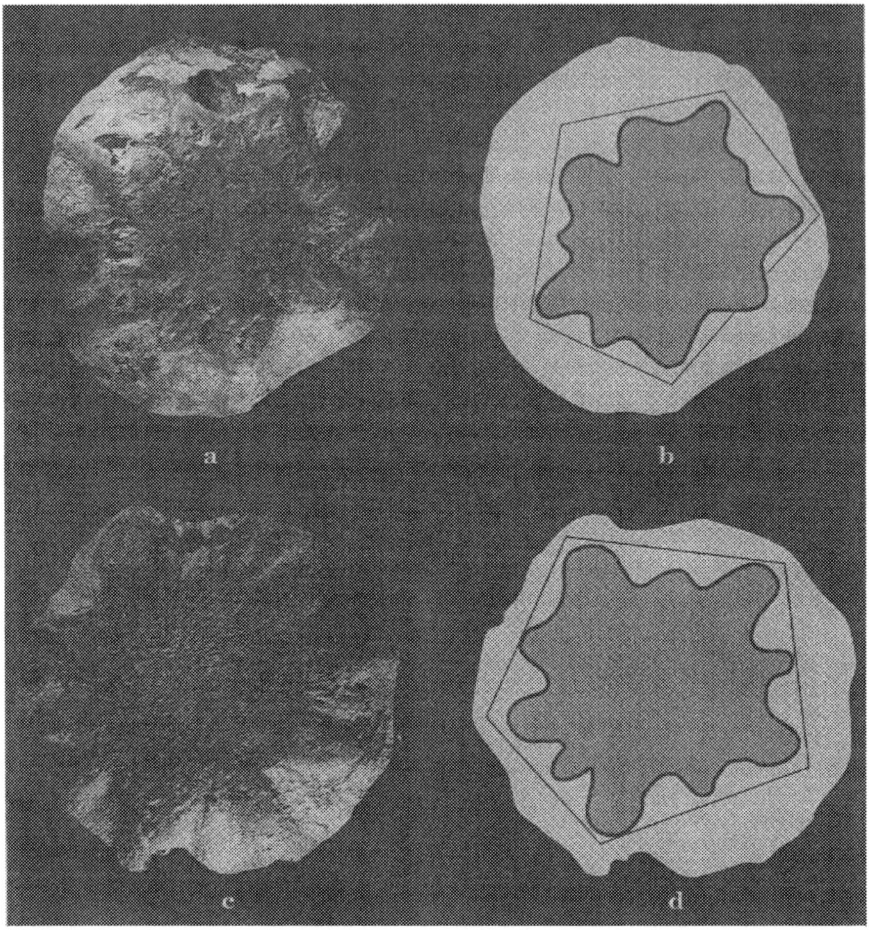

Figure 11. Orange-colored fruit-end fragments from Zone B of Guilá Naquitz that exhibit pentagonal major-minor ridge morphology in peduncle scars and associated fruit-end lobing diagnostic for *C. pepo*. (a–b) Fruit-end fragment from Square C11, Zone B1 (INAH 42), maximum basal diameter 20.2 mm., 6115 cal. B.C. (c–d) Fruit-end fragment from Square B9, Zone B (INAH 58), maximum basal diameter 23.6 mm., 5805 cal. B.C. (See Tables 2, 4).

peduncle cross-section. In addition, they help establish the identity of the domesticated squash as *C. pepo* ssp. *pepo*, based on their 10-ridge lobing pattern.

These fruit-end fragments also document two other changes in the Guilá Naquitz *C. pepo* squash that indicate domestication and deliberate human selection. The first of these changes involves a fruit form or shape that differs substantially from the wild *Cucurbita* morphotype profile. In contrast to wild *Cucurbita* gourds, which have a generally spherical shape with some variation toward oblate, pronate, and pyriform fruit forms, the two peduncle-end fruit fragments shown in Figure 11 angle abruptly down and away from the peduncle attachment in a manner similar to that of modern "zucchini" cultivar forms of *C. pepo* ssp. *pepo.* Although it is difficult to determine what the overall shape of these fruits was, beyond the small peduncle-end fragments that were recovered, the size and curvature of several of the other rind fragments from the cave suggest that, rather than having an elongated, tubular, zucchini shape, the fruits may have expanded out into a more globular form.

The other remarkable characteristic that differentiates the Guilá Naquitz domesticated squash from the wild morphotype profile is its color. Whereas modern wild *Cucurbita* gourds almost always range in color from solid white or ivory to light green to white with green stripes (Robinson and Decker-Walters 1997:94), two of the Guilá Naquitz peduncle-end fruit fragments (Figure 11) retain a vibrant pumpkin-orange color, which is a hallmark of the Mexican lineage of *C. pepo* ssp. *pepo.*[2] Beneath this orange surface layer, the peduncle-end fruit-rind fragment shown in Figure 11c contained spherical phytoliths 70 to 90 μm in diameter, which appear to be scalloped (Figure 12), comparable to those proposed as being diagnostic for the genus *Cucurbita* (Bozarth, 1987; Piperno, 1988:63, 226).

Turning to a consideration of the stratigraphic context of the 18 *Cucurbita* peduncles and peduncle-end fruit fragments from the preceramic zones of Guilá Naquitz cave, a clear pattern can be seen that parallels that already observed for the *Cucurbita* rind fragments in the cave. The 10 peduncles and peduncle scars that fall within the size and morphological parameters of wild *Cucurbita* gourds, along with thin *Cucurbita* rind fragments, occur throughout Zones D, C, and B1–B3 of the cave, reflecting the long-term use of wild *Cucurbita* gourds (Figures 7, 9). In contrast, six of the eight peduncles and peduncle scars that fall outside the wild *Cucurbita* morphotype profile in terms of size and shape were recovered from Zones B1–B3 contexts and parallel the documented Zones B1–B3 increase in rind thickness that reflects the appearance of a new, domesticated fruit form.

The only two specimens that fall outside the wild morphotype profile and that are not from Zones B1–B3 are both from Zone C. The first of these, which is noticeably shriveled, has a pentagonal base, a maximum basal diameter of 20.8 mm (Table 4), and if not intrusive from Zones B1–B3 indicates the presence of a domesticated fruit form in a pre-Zone B3 temporal context. The second Zone C specimen is also interesting in that it falls just outside the wild morphotype profile in terms of both size and morphology and may represent an early stage of human selection for a modified *Cucurbita* fruit form. Measuring 13.0 mm in maximum diameter, the peduncle scar on this fruit-end fragment is decidedly pentagonal rather than circular in shape. In addition, whereas the peduncle scar of this fruit-end fragment is depressed, suggesting an oblate fruit form similar to that of

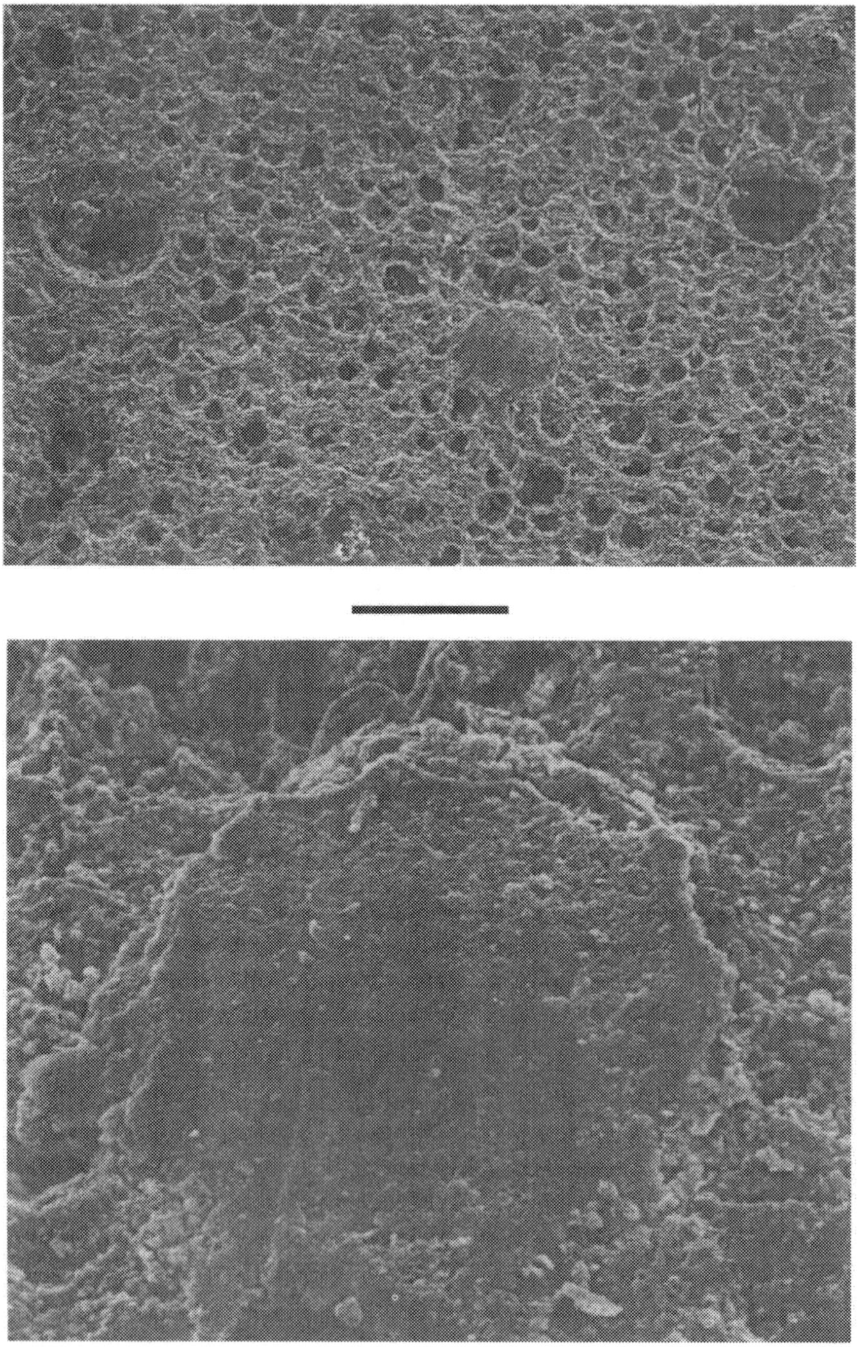

the wild morphotype specimens in the Guilá Naquitz assemblage, it differs from them in having 10 small bulges or lobes arrayed around the peduncle scar.

These two Zone C specimens might indicate pre-Zone B3 human selection for a modified fruit form in *Cucurbita*. Direct AMS dates were obtained for samples of the three Zones B1–B3 specimens that had the largest measurements for basal peduncle diameter and that could confidently be identified on the basis of morphology as *C. pepo* ssp. *pepo*. The AMS dates (6165, 6115, 5805 cal. B.C.) obtained for these three specimens (Tables 2, 4), along with their horizontal provenience (Squares B9, C11), place them squarely within the Zones B1–B3 (6480–5805 cal. B.C.) spatial-temporal occupational cluster located toward the rear of Guilá Naquitz cave (Figure 4) and clearly indicate that by c. 6000 B.C. deliberate human selection of cultivated *Cucurbita pepo* ssp. *pepo* had produced a fruit distinctly different from the wild *Cucurbita* morphotype. The peduncle of this new fruit form is twice as large, in terms of basal diameter, as wild *Cucurbita* gourds, has a pronounced pentagonal, as opposed to round, attachment scar, and also exhibits the distinctive alternating large and small 10-ridge morphology diagnostic for domesticated *C. pepo*. This alternating large and small 10-ridge stem morphology extends down onto the fruit, which also changes dramatically, both in shape, and coloration. In contrast to wild *Cucurbita* gourds—small, spherical to pyriform in shape, and white to white and green striped in color—two of the Guilá Naquitz Zones B1–B3 fruits are orange in color, and these two orange-colored peduncle-end fruit fragments indicate that the distinctively ridged or lobed fruit angled steeply away from the peduncle scar in a zucchini-like manner. In addition to demonstrating the presence, by 6000 B.C., of a domesticated *C. pepo* ssp. *pepo* squash considerably changed from the wild morphotype, the three AMS-dated Zones B1–B3 specimens from Guilá Naquitz may also document an ongoing process of selection for an increase in fruit size, judging from the temporal trend in maximum basal peduncle diameter shown in Figure 13.

Cucurbita Seeds

In their analysis of the cucurbits from preceramic levels of Guilá Naquitz, Whitaker and Cutler identified 17 seeds as assignable to the genus *Cucurbita*. Unfortunately, only 10 of these 17 seeds are present in the INAH collections (Table 5).[3] The three obvious and interrelated questions to be asked of this small collection

Figure 12. Scanning electron micrographs showing opal phytolith features on the surface of a *Cucurbita pepo* fruit-end fragment from Guilá Naquitz (INAH 58, Tables 2, 4). (above) Rind surface showing two "craters" where phytoliths have been lost, along with an intact phytolith. Craters and intact phytolith are 70-90 microns in diameter. Scale bar = 100 μm. (below) Closeup of intact phytolith shown in a. Although surface morphology of the phytolith is obscured, it appears to have the scalloped form that has been described for modern *Cucurbita* rind phytoliths (note scalloped edge of phytolith). Scale bar = 10 μm.

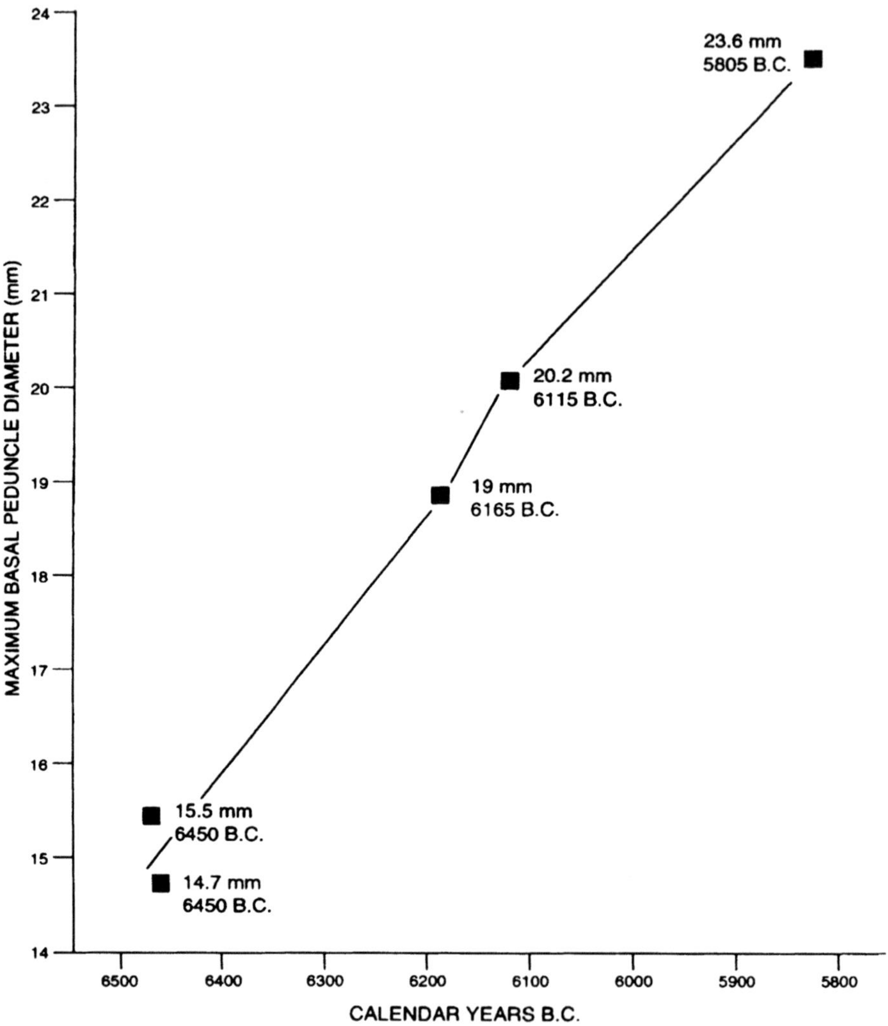

Figure 13. Increase in the size of peduncles of domesticated *Cucurbita pepo* in preceramic habitation zones of Guilá Naquitz cave between 6500 and 5800 cal. B.C. The two peduncles assigned an age, by association, of 6450 cal B.C. come from Square C11, Zone B1, which also yielded seeds with AMS dates of 6480, 6475, 6465, 6425 cal. B.C. (Table 2).

of preceramic *Cucurbita* seeds involve their taxonomic assignment, their status as wild or domesticated plants, and their temporal and spatial context in the occupational history of Guilá Naquitz.

Whitaker and Culter identified 8 of 17 seeds as representing a local species of wild *Cucurbita* gourd and referred to them as *Cucurbita* sp. (Whitaker and Cutler, 1986:275, 277–278, Tables 20.2–20.8). The other nine *Cucurbita* seeds were identified as representing a domesticated species of squash—*Cucurbita pepo*. Whitaker and Cutler, however, did not offer any discussion of the criteria they employed to distinguish the seeds of wild *Cucurbita* gourds from those of domesticated *C. pepo*, nor did they specify the morphological characteristics they used to identify the domesticate seeds as representing *C. pepo* rather than other domesticated species such as *C. moschata* or *C. argyrosperma*. They did, however, discuss the presence of *C. pepo* and the absence of these other two squash species in terms of the environmental setting of the cave:

> The dominance of *C. pepo* at Guilá Naquitz is most likely explained by the distinct ecological preferences of the cultivated species of *Cucurbita*. One would normally expect to find *C. mixta* [*C. argyrosperma*] and *C. moschata* cultivated at the latitude of the cave. . . . The cave, however, is located at an altitude of about 1926 m . . . in an arid region. . . . The days are predominantly warm, with cool nights and minimum moisture for plant growth. Such conditions are well suited to cultivars of *C. pepo*. On the other hand, most cultivars of *C. moschata* and *C. mixta* [*C. argyrosperma*] usually grown at this latitude thrive under high temperatures and moist, humid conditions, but they do not tolerate cool temperatures. (Whitaker and Cutler, 1986:276)

In addition to this environmental explanation for the absence of *C. argyrosperma* and *C. moschata* in the preceramic zones of Guilá Naquitz, consideration of the morphology of the 10 *Cucurbita* seeds shows them to be more similar to the seeds of modern *C. pepo* than they are to those of either of the other two species. Over the past half-century, a number of researchers have addressed

Table 5. *Cucurbita* Seeds from Guilá Naquitz

INAH catalog/ sample number	Square/ Zone	Length (mm)	Width (mm)	Width/ length ratio	Age: Dendro-calibrated intercept
14 (GN 22)	E9/C	13.8	7.4	.53	7975 B.C.
23 (GN 24)	E11/B(B1)	13.2	8.8	.66	8025 B.C.
5 (GN 21)	C11/B2	12.5	7.9	.63	6475 B.C.
5	E7/B3	12.3	7.2	.58	
7 (GN B)	C11/B1	12.1	7.3	.60	6465 B.C.
7 (GN D)	C11/B1	12.0	7.0	.58	6425 B.C.
7 (GN A)	C11/B1	11.4	7.2	.63	6480 B.C.
1	F9/C	11.3	7.5	.66	
15 (GN 23)	E7/C	10.7	7.0	.65	
10	D8/B2	Fragmented			

the difficult challenge of correctly and consistently distinguishing between the seeds of different domesticated species of *Cucurbita*. A variety of potentially diagnostic morphological characteristics has been proposed, including seed dimensions (e.g., length, width, width/length ratio, margin width, seed outline) and numerous nonmetrical variations in the seed scar (hilum), seed body, and seed margin (Burgess-Terrel, 1979; Cutler and Whitaker 1956, 1961; Decker, 1984, 1986:70; Decker and Newsom, 1988; Decker and Wilson, 1986; Russell, 1924; Tapley et al., 1937; Whitaker and Bohn, 1950). At various times, taxonomic authorities have also cautioned about the potential difficulties involved in accurate assignment of seeds to particular species (Cutler and Whitaker, 1956, 1961:474-475; Heiser, 1985:12; Nee, 1990; Whitaker, 1981:462). Whitaker, for example, stated that the seeds of *C. pepo* and *C. moschata* "are oftentimes difficult to separate without a great deal of experience" (Whitaker, 1981:462), whereas Nee similarly indicated that "*C. moschata* is often difficult to distinguish from *C. argyrosperma*. . . . [Their] seeds are often very similar" (Nee, 1990:63).

The difficulties inherent in attempting to distinguish between the seeds of these three modern domesticated species of squash are further compounded in regard to the Guilá Naquitz seeds by the substantial temporal and developmental gap that separates the preceramic specimens from present-day *Cucurbita* cultivars. These modern squashes all have long and rich crop histories of substantial human selection and directed diversification, much of which remains to be documented in the archaeological record. As a result, the Guilá Naquitz seeds should not be expected to provide an exact match—to incorporate all the morphological attributes present in the seeds of any one of the present-day *Cucurbita* domesticates. Given the considerable amount of time for human selection to have occurred since the preceramic-period seeds were deposited in Guilá Naquitz cave, it is surprising that they provide such a close match to present-day *Cucurbita pepo*.

Take, for example, the single criterion that today is the most obvious and most frequently mentioned way of distinguishing between the seeds of *C. argyrosperma*, *C. moschata*, and *C. pepo*—the coloration of the seed margin relative to the seed body. With few exceptions, *C. moschata* and *C. argyrosperma* seeds have a margin different in color from the body—usually darker (Cutler and Whitaker, 1961:478; Decker, 1986:70; Merrick and Bates, 1989:95-97; Nee, 1990:63). The body and margin of *C. pepo* seeds, in contrast, are the same color—uniformly off white to pale tan in color (Cutler and Whitaker, 1961:478; Decker, 1986:70; Nee, 1990:59). Although seed body/seed margin coloration is in most cases not very applicable to the analysis of *Cucurbita* seeds from archaeological contexts because of preservation problems, the *Cucurbita* seeds and fruit-end fragments from Guilá Naquitz exhibit remarkable retention of color (Smith, 1997a:Figure 2). All the Guilá Naquitz seeds in the INAH collections (Table 5) have seed bodies and seed margins that are uniformly light tan in color, comparable to modern *C. pepo* seeds.

The diagnostic variation in seed body and margin coloration observed in modern *C. pepo, C. moschata, and C. argyrosperma* also is linked to anatomical differences in the seeds of these three species. In both *C. argyrosperma* and *C. moschata,* seed bodies have thick surface layers of cells that are soft and pulpy, often white in color, and sometimes exhibiting irregular patterns of cracks or fissures. Such surface crack patterns are more common on the surface layer of *C. argyrosperma*. When viewed in cross-section, the surface layer of seeds of this species can be seen to be composed of long, vertically arranged cells that when separated can form deep cracks. In *C. moschata,* this soft white surface layer is both thinner than and different from *C. argyrosperma* in cross-section structure. The seeds of *C. pepo*, in contrast, entirely lack this soft, usually white surface layer, and the seed body is typically hard and "smooth" (a characteristic lightly bumpy surface texture is evident under magnification). All the Guilá Naquitz *Cucurbita* seeds lack a soft white surface layer and have a hard seed body with a slightly bumpy surface texture comparable to that of modern *C. pepo* seeds. The Guilá Naquitz seeds are also similar to modern *C. pepo* seeds and different, in general, from those of the other two species under consideration in that they all exhibit a prominent and well-rounded ridge that separates the body of the seed from the seed margin (Figure 14).

Seeds of the three present-day squash species being discussed also exhibit clear differences in their margin morphology. All three species have an overlying margin layer of the seed variously described as "stringy" (Decker 1986:70) or consisting of "threads" (Cutler and Whitaker 1961:478) or "hairs." This overlying layer of hairs/threads/strings begins or is "rooted" along the outer edge of the seed body or along the inner edge of the ridge (if present) that separates the seed body from the margin. It extends up and over the ridge and out onto the margin of the seed. Made up of a large number of individual threadlike structures that run parallel to one another and are loosely to strongly "cemented" together, this overlying margin thread layer exhibits considerable morphological variation between and within species.

In *C. argyrosperma*, for example, the overlying thread layer in most modern cultivar varieties extends out onto the margin at least as far as the midline of the margin, where the thread layers from both sides of the seed converge. In some cultivars, they overlap for a short distance, one on top of the other, and in other cultivars, the thread layers meet and flair outward together along the margin midline. When the thread layers flair outward in this manner, there is often an associated small underlying margin midline ridge. In both the overlapping and midline flaring patterns of morphology of seed margin thread layers, seed margin color is provided by the varnishlike sheen of the thread layer itself. Underneath the thread layer, the seed margin is soft and irregular in surface form and often lighter in color than the thread layer. In other modern cultivar varieties of *C. argyrosperma*, however, the overlying margin thread layer does not extend very far at all, and the

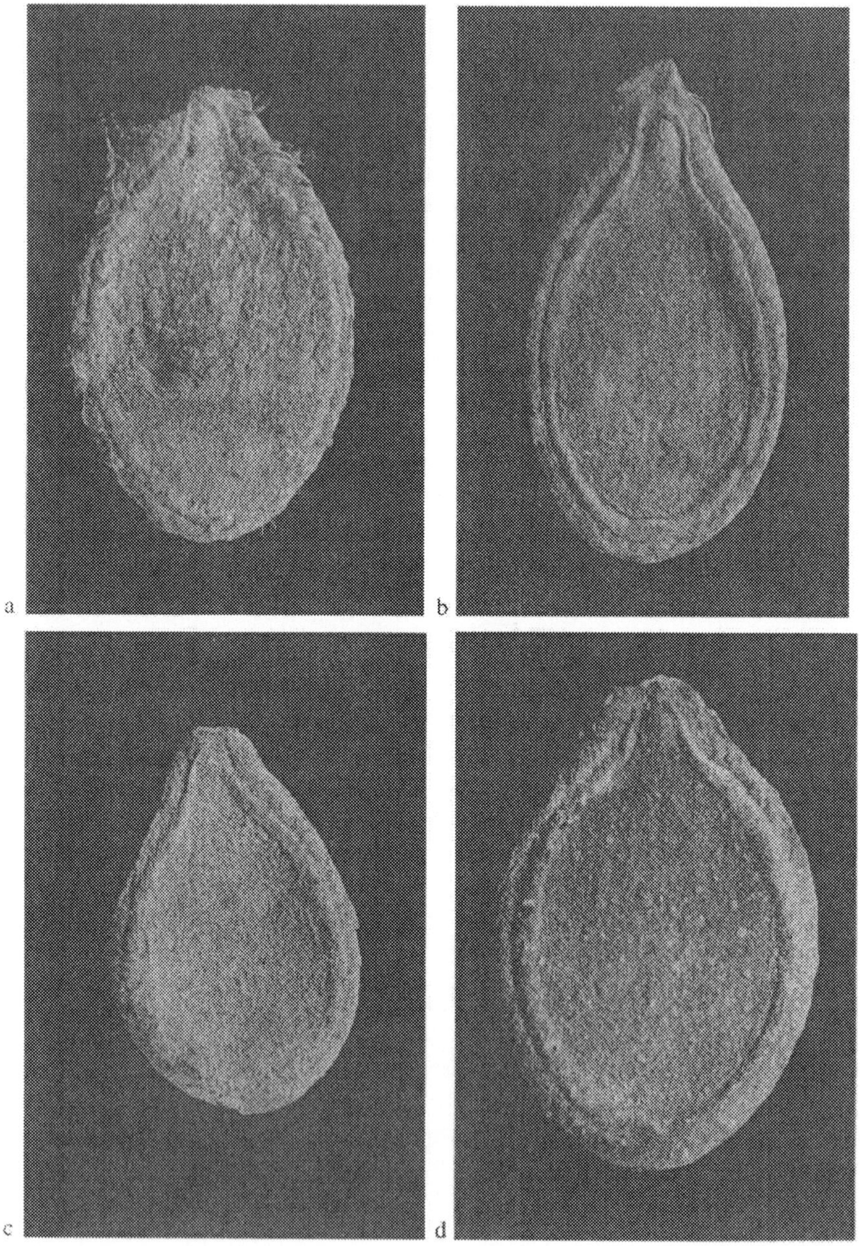

Figure 14. *Cucurbita pepo* seeds from Guilá Naquitz cave yielding AMS dates in the Early Archaic. (a) Seed from Square C11, Zone B2 (INAH 5), measuring 12.5 mm in length and yielding an AMS ^{14}C date of 6475 cal. B.C. (b) Seed from Square E9, Zone C (INAH 14), measuring 13.8 mm in length

seed margin's distinctive green color comes not from the thread layer, but from the exposed or "naked" margin itself, which is typically firm, substantially swollen, and irregularly rounded in shape.

In comparison to *C. argyrosperma,* which exhibits considerable intraspecies variation in margin thread layer characteristics, both *C. pepo* and *C. moschata* display relatively consistent margin thread layer morphology. In most modern cultivar varieties of *C. pepo*, for example, seeds have a margin thread layer that extends only a short distance out onto the seed margin. The largely exposed or "naked" seed margin is smooth, hard, uniformly rounded, and the same color as the seed body. The margin thread layer in modern cultivar varieties of *C. moschata*, in contrast, appears to be a single unbroken structure that extends completely around the seed margin. Usually having a reddish to gold varnishlike sheen, this thread layer has a wavy convoluted appearance with crinkles that run along the margin. These crinkle patterns are often impressed into the soft undulating surface of the underlying margin surface.

Although these differences in margin thread layer morphology work well, for the most part, in distinguishing among the seeds of modern *C. pepo, C. moschata,* and *C. argyrosperma*, they are not as useful in sorting seeds from archaeological contexts. Seed margin thread layers are susceptible to abrasion, particularly along the margin midline. Such abrasion can break, fray, and separate the margin threads of all three species, making taxonomic assignment on the basis of this character difficult. The Guilá Naquitz seeds, however, are generally in good condition. Several seeds have a short thread *C. pepo*-like appearance (INAH 7, 14, 15; Table 5), whereas others (INAH 5, 23) have a margin thread layer that completely covers the margin, similar to modern *C. moschata*. Complete coverage of the seed margin by a thread layer is not a characteristic that occurs only in *C. moschata*, however. It is occasionally also present in seeds of modern cultivars of *C. pepo*.[4] The margin of the Guilá Naquitz seeds also is somewhat bumpy and differs from the smooth appearance of naked margin seeds of modern *C. pepo*. Thus the Guilá Naquitz seeds cannot be said to provide an exact match with modern cultivar varieties of *C. pepo* in all respects. When all the seed body, ridge, and margin attributes discussed above are considered together, however, the Guilá Naquitz seeds are clearly much more similar in appearance to *C. pepo* than to the other two species considered. On the one hand, they exhibit a number of seed characters that do not occur in modern *C. argyrosperma* or *C. moschata* (e.g., uniform color, prominent ridge, absence of seed-body surface-cell layer, short margin hairs, and naked margin). At the same time, however, they do not exhibit

---

and yielding an AMS ^{14}C date of 7975 cal B.C. (note that the other side of this seed is illustrated in Smith, 1997a). (c) Seed from Square E7, Zone C (INAH 15), measuring 10.7 mm in length. (d) Seed from Square E11, Zone B (INAH 23), measuring 13.2 mm in length and yielding an AMS ^{14}C date of 8025 cal. B.C. (Tables 2, 5).

any morphological characteristics that cannot also be found, if only occasionally, in modern cultivars of *C. pepo*.

Given that a strong case can be made for assigning the Guilá Naquitz seeds to *C. pepo*, the next question to be addressed is whether they represent, in this species category, a wild or a domesticated plant. In a number of articles published in the 1960s and 1970s, Harlan, de Wet, and Price outlined, for the first time, a clear set of morphological criteria for distinguishing between wild and domesticated seed plants on the basis of changes in reproductive propagules (seeds and fruits), which reflect deliberate and sustained human storage, planting, and harvesting of seed stock over a number of years (de Wet and Harlan, 1975; Harlan and de Wet, 1965; Harlan et al., 1973). These key morphological changes, which mark responses by the now managed plants to new sets of seedbed and harvesting selective pressures, are automatic and independent of any deliberate or directed selection on the part of humans and collectively have been termed "the adaptive syndrome of domestication" (see Smith, 1992, 1998, for a fuller discussion of this topic). Many of the automatic morphological changes in seed plants that occur as the result of deliberate human planting and harvesting, however, are often difficult to document in archaeobotanical assemblages. One of the most obvious and most archaeologically observable of these changes that mark initial domestication is an increase in seed size in response to seedbed selective pressure.

In their analysis of the Guilá Naquitz *Cucurbita* seeds, Whitaker and Cutler did not identify the criteria they employed to distinguish between seeds of wild *Cucurbita* gourds and those of domesticated plants, but seed size was clearly not used. They did note an increase in the size of *C. pepo* seeds up through the preceramic levels of the cave and "suggest there is a trend toward selection, perhaps largely unconscious, by the earlier cultivators for larger size seed and presumably larger fruit" (Whitaker and Cutler, 1986:276, Table 20.1). But seed dimensions were not used to distinguish between the smaller seeds of wild gourds and the larger seeds of domesticated squash. The two seeds identified as wild *Cucurbita* sp. in their Figure 20.2, for example, both appear to be larger than all but three of the nine seeds identified as representing domesticated *C. pepo* (Whitaker and Cutler, 1986:276–277, Table 20.1)

In the current study, to consider seed size as a criterion for differentiating between the seeds of wild and domesticated *C. pepo* in the Guilá Naquitz assemblage, it was first necessary to establish a wild *Cucurbita* standard or baseline of comparison. In theory, the best source of wild baseline seed-size measurements would consist of a large and representative sample of seeds from the wild *Cucurbita* gourd that grew in the area around Guilá Naquitz while it was occupied and that was the wild ancestor of the Mexican lineage of domesticated *C. pepo*. Unfortunately, although *C. sororia* grows abundantly today in the Valley of Oaxaca, it is not the progenitor of the Mexican lineage of domesticated *C. pepo,* which has yet to be identified, collected, or described. In the absence of this ideal wild gourd baseline of comparison, the next best alternative was to compile seed-size statistics from as many different species of wild *Cucurbita* gourds as possible and to use them to construct a general wild *Cucurbita* gourd morphotype profile that encompasses as much variation as can be documented in wild populations and that in turn can be employed as a baseline of comparison for consideration of the wild versus domesticated status of the Guilá Naquitz seeds. To construct such a general wild *Cucurbita* gourd seed profile, 717 seeds from 63 separate accessions, representing 12 different modern species, subspecies, or varieties of wild *Cucurbita* gourds, were measured (Figure 15). With a few exceptions (e.g., *C. pedatifolia, C. galeottii*), all known North American (Mexican and U.S.) wild *Cucurbita* species are included in Figure 15 (see Nee, 1990:58). The number of seeds measured for some species, however, is admittedly small. In addition to the 12 modern taxa included in Figure 15, summary seed-length information is also presented for the Page Ladsen seed assemblage: a collection of 35 late Pleistocene (c. 10,500 ^{14}C years B.C.) seeds of a wild *Cucurbita* gourd recovered from American mastodon (*Mammut americanum*) dung deposits in north Florida (Newsom et al., 1993). The mean seed length values for these 13 wild *Cucurbita* gourd taxa range in value from a low of 8.0 mm (*C. fraterna*) to a high of 10.7 mm (*C. gracilior*). The smallest of the 717 seeds measured was 7.0 mm long, while the largest was 11.9 mm in length. Most important for the current analysis, although the largest seeds of the vast majority of the reference taxa (10 of 13) exceeded 10.0 mm in length, none was more than 12.0 mm long (24 of the 717 seeds were longer than 11.0 mm, and only 4 of these measured more than 11.5 mm in length). Taken together, these maximum length values provide a strong basis for defining the upper size-range ceiling for seeds of North American (Mexican and U.S.) wild *Cucurbita* gourds at 10.0–12.0 mm in length. One can argue, alternatively, that the three *C. pepo* taxa in this reference class of 13 constitute a more appropriate baseline of comparison for the Guilá Naquitz seeds and that their upper limit for seed length (c. 11.0 mm) provides a better wild-domesticated boundary line than the c. 12.0 mm seed-length ceiling derived from the larger reference class. After all, the three *C. pepo* gourds in question are all taxonomically close to the Guilá Naquitz specimens, and one of the three (*C. pepo* ssp. *ovifera* var. *ozarkana*) has been identified as the wild progenitor of the eastern lineage of domesticated *C. pepo* squashes (Decker-Walters et al., 1993). Robinson and Decker-Walters (1997:7) and others have pointed out, however, that the seeds and fruits of the modern cultivars of the eastern North American lineage of *C. pepo* are smaller than those of the Mexican lineage; it is possible that this size difference may also exist in the wild ancestor populations. As a result, although 11.0 mm seems to provide a good minimum baseline-length value for identifying domesticated *C. pepo* seeds in eastern North American archaeobotanical assemblages, the larger reference class value of 12.0 mm provides a more prudent minimum seed-length value for the designation of domesticated *C. pepo* in Mesoamerican contexts, at

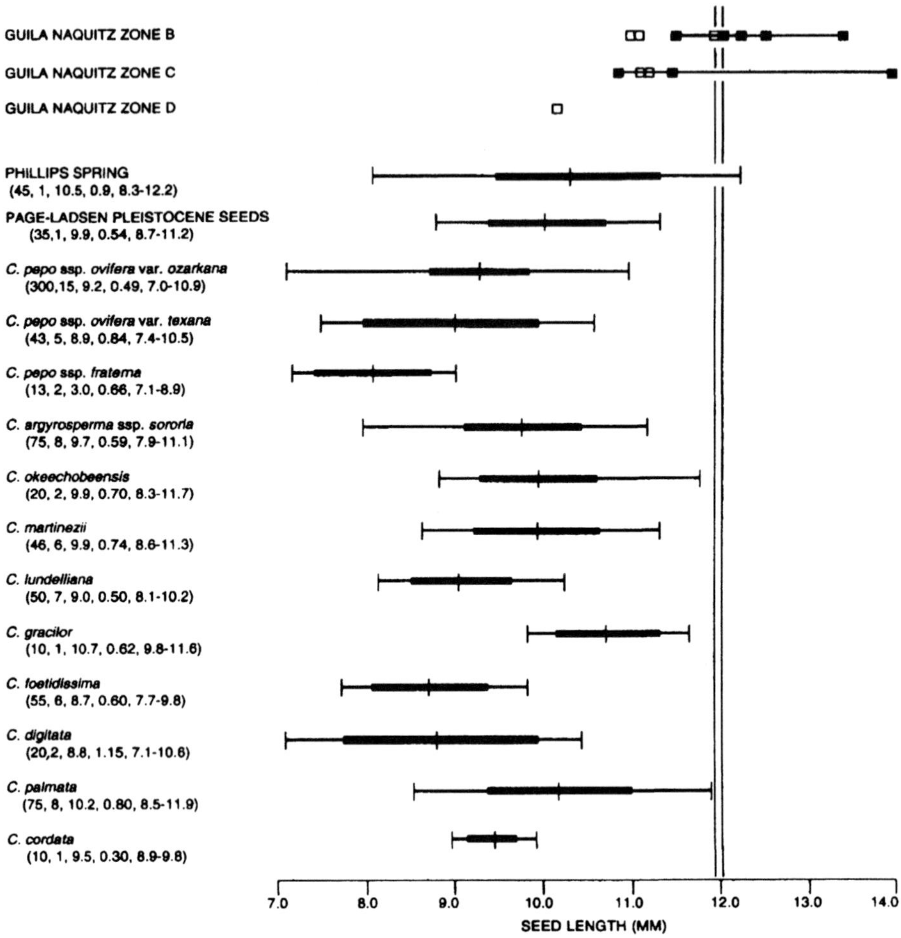

Figure 15. The size of *Cucurbita pepo* seeds from preceramic habitation zones of Guilá Naquitz cave, compared with *Cucurbita* seeds from the Pleistocene Page-Ladsen site and 12 modern taxa of wild *Cucurbita* gourds. Measurements for the Guilá Naquitz seeds are given in Table 5. Statistics given for the 13 reference taxa are number of seeds, number of accessions, mean, standard deviation, and range. Based on seed-length measurements obtained from cataloged specimens from the following museum and herbarium collections—Fairchild Tropical Garden: DDW M10, 658, 659, 747, 750, 751; Field Museum of Natural History: 51359, 90365, 108583, 436259, 1883666, 1972892; Illinois State Museum: AZAC 33, 651, 671, 678, 679, 681, 682, 684, 685, 686, 688; Missouri Botanical Garden: 778804, 1811607, 2323729, 2393083, 2393107, 2407657, 2407661, 2407662, 2483791, 3264716, 3379643, 3379647, 3534773; National Museum of Natural History: 64648; Research collections of D. Decker-Walters, B. Smith.

least until the wild progenitor of the Mexican *C. pepo* lineage is identified and described.

When the nine measurable seeds from Zones B1–D of Guilá Naquitz now in the INAH collections are viewed against the comparative backdrop provided by this reference class of wild *Cucurbita* gourds, a pattern emerges that parallels those already documented for *Cucurbita* rind and peduncle assemblages from the cave. This pattern is even clearer when the six seeds that were measured, and in some cases illustrated, by Whitaker and Cutler, but that did not make it into the INAH collections, are also considered (represented by open squares in Figure 15). When these seeds are included, a total of nine seeds (one from Zone D, four from Zone C, four from Zones B1–B3) fall below the 12.0 mm seed-length boundary line, suggesting the continuing use of wild *Cucurbita* gourds throughout the occupation of the cave and providing further confirmation of the pattern already documented in both rind (Figure 7) and peduncle (Figure 9) assemblages. This group of nine Guilá Naquitz seeds that fall into the size range of wild *Cucurbita* gourds and that, as a result, cannot be clearly identified as domesticated *C. pepo* squash includes the 10.0-mm-long seed from Zone D identified in the initial analysis as domesticated *C. pepo* (Whitaker and Cutler, 1986:275–276).

In contrast, a total of six seeds, all of which are in the INAH collections, fall at or above the 12.0-mm minimum length boundary line for distinguishing between the seeds of wild and domesticated *Cucurbita pepo* in the Guilá Naquitz assemblage (Table 5). Samples were taken from the seed body of five of the six seeds that fall above the 12.0 mm boundary, along with one that is smaller than 12.0-mm, for AMS ^{14}C age determinations. Four of the six dated seeds formed a tight spatial-temporal grouping within the Zones B1–B3 cluster defined above (Table 2). All four of these seeds (11.4, 12.0, 12.1, and 12.5 mm in length) were recovered from Zone B1, excavation unit C11, and they ranged in age from 6480 to 6425 cal. B.C. (Tables 2, 5). Given this tight spatial and temporal grouping, these four seeds may well have been deposited at the same time (see discussion of pattern 11, Spencer and Flannery, 1986:367) during an occupational episode that was separate from, and earlier than, the Zone B1 occupation that yielded the three AMS-dated fruit-end fragments and peduncle of domesticated *C. pepo* squash discussed above (Tables 2, 4). Two other fruit-fragment specimens of domesticated *C. pepo* squash as yet not directly dated could well go with these four Zone B1 occupational cluster seeds that date to ca. 6500 cal. B.C. A distinctively lobed *C. pepo* peduncle having a maximum basal diameter of 14.7 mm and a fruit-end fragment with a 15.5 mm pentangular-shaped peduncle scar were recovered from the same location as the four AMS-dated seeds (Zone B1, excavation unit C11) (Table 4, Figure 13).

The other two dated seeds, which at 13.2 and 13.8 mm in length were the largest *C. pepo* seeds in the Guilá Naquitz assemblage, also formed a tight spatial-temporal group. These two seeds fell within the much earlier Zones C–D Early Archaic occupation cluster (Tables 2, 5). Recovered from Zone B1, Square E11, and Zone C, Square E9, the two seeds in question were very close in age (8025, 7975 cal. B.C.), and both could well have been deposited during the same brief occupational episode. At 13.2 and 13.8 mm in length, these two seeds are 10% and 15% larger, respectively, than the 12.0-mm-length ceiling defined for wild *Cucurbita* gourds and provide strong evidence for the cultivation of a domesticated *C. pepo* ssp. *pepo* squash by the inhabitants of Guilá Naquitz by about 8000 B.C. These two remarkably well-preserved seeds, which had lain buried for 10 millennia in a small and shallow cave in Oaxaca, Mexico, until carefully excavated by Kent Flannery in spring 1966, in fact provide the earliest evidence yet available in the Americas for plant domestication and the beginning of the transition from hunting and gathering to food production and eventually to agriculture.

It is important, however, to reiterate that the case for domesticated *C. pepo* ssp. *pepo* in the preceramic occupation zones of Guilá Naquitz does not rest solely on these two seeds and their relative size and temporal context. They should not be viewed in isolation, but rather be recognized as making up one of a number of intertwined and mutually reinforcing lines of evidence that, when drawn together, provide a clear if only partial view of the early evolutionary history of this domesticate.

THE INITIAL DOMESTICATION OF *CUCURBITA PEPO*

As is the case with a number of other New World domesticates (e.g., *Phaseolus, Chenopodium*; see Smith, 1995a:146–181), *C. pepo* was independently domesticated in different regions, at different times, by different human societies, from geographically separated wild progenitor populations. Although Mexico had long been identified as the location where *C. pepo* was first brought under domestication, several scholars, including C. Wesley Cowan and Charles Heiser, proposed in the early 1980s that this squash species may also have been independently domesticated in the eastern woodlands of the United States from a wild *Cucurbita* gourd that was at that time unrecognized but believed to be indigenous to the region. In her dissertation research of the mid-1980s on the taxonomy of *C. pepo*, Decker (1986) provided the first strong empirical evidence to support this dual-domestication theory. Her allozyme analysis revealed a major division of modern cultivar varieties of *C. pepo* into two developmental lineages—*C. pepo* ssp. *pepo*, which includes all the orange-colored Mexican land races of pumpkins, and *C. pepo* ssp. *ovifera*, which includes the green- and yellow-colored acorn and summer squashes. Decker concluded that this division of *C. pepo* into two major lineages was very likely rooted in its independent domestication in Mexico and the eastern United States. Subsequent field and laboratory research in the early 1990s confirmed Decker's theory. A wild *Cucurbita* gourd (*Cucurbita pepo* ssp. *ovifera* var. *ozarkana)* was recognized in the eastern United States and described (Cowan and Smith, 1993) and was subsequently identified as the wild progenitor of the eastern U.S. lineage of domesticated *C. pepo* squash (Decker-Walters et al., 1993). Although the wild ancestor of the Mexican lineage of *C. pepo* has not yet been identified, archaeobotanical assemblages that cast at least some light on the initial domestication of this species are available from both Mexico and the eastern United States.

When the early developmental history of domesticated *C. pepo* in these two regions is compared, based on available archaeobotanical information, a number of general parallels can be seen, sometimes finding expression in different ways. These developmental similarities in turn compare closely to theoretical expectations in terms of both the adaptive syndrome of domestication and patterns of directed human selection. In both regions, for example, the transformation of a wild *Cucurbita* progenitor into a domesticated crop plant began with a key escalation of human intervention in the life cycle of wild populations of a *Cucurbita pepo* gourd—the deliberate storage and planting of seed stock in prepared seedbeds. In so doing, human planters inadvertently established a new arena of competition for emerging seedlings—a new set of "seedbed selective pressures"—that favored those plants that could sprout and grow the fastest, literally shading out their competitors during the growing season. Under this new set of rules for reproductive success, those seeds with reduced germination dormancy and greater start-up energy reserves (contained in a larger seed) would have the competitive edge and a higher probability of surviving to harvest time and making a substantial contribution to the seed stock set aside for next year's planting. An increase in seed size above that seen in wild progenitor populations, and in wild *Cucurbita* gourds in general, would as a result be an automatic response by plants that were part of early cultivated plots of *Cucurbita pepo* and could be expected, as part of the adaptive syndrome of domestication, to appear unbidden and quite early in the ongoing transformation of a small, thin, hard-shelled wild gourd into a domesticated squash. Other morphological changes reflecting deliberate human selection of desired traits would appear only later in the evolution of the crop plant.

There is, interestingly enough, evidence for this developmental sequence—of seed size increase, an automatic response to seedbed selective pressure, preceding any other morphological changes, either automatic or intentionally induced—in the initial domestication of the eastern United States lineage of *C. pepo* from progenitor populations of the wild Ozark gourd (*C. pepo* ssp. *ovifera* var. *ozarkana*). Although thin rind fragments of wild ancestor *Cucurbita* gourds, along with occasional small seeds, have been recovered from a number of mid-Holocene (8000–4000 B.P.) contexts in the eastern United States (see Cowan, 1997; Smith, 1992), the earliest evidence for domesticated *C. pepo* ssp. *ovifera* in the region comes from the Phillips Spring site in west-central Missouri (King, 1985).

Dating to c. 3000 cal. B.C.,[5] the *Cucurbita* assemblage from Phillips Spring included more than 50 *C. pepo* seeds, a dozen rind fragments, and four fruit-end fragments bearing peduncle scars.

Generally accepted as representing an early stage of domesticated *C. pepo* in the eastern United States, the Phillips Spring seed assemblage has a mean length value larger than all but one of the 13 wild *Cucurbita* gourd reference taxa listed in Figure 15. Although only two Phillips Spring seeds meet or exceed the 12.0-mm seed-length boundary line employed in the Guilá Naquitz analysis, a full third of the Phillips Spring assemblage (15 of 45 measured seeds; Figure 15) falls above the 11.0-mm wild seed-length ceiling proposed for the eastern United States (Cowan and Smith, 1993:43) and thus provides strong evidence for deliberate planting, an adaptive response to seedbed selective pressure, and the domestication of *C. pepo* var. *ovifera.* The rind and fruit-end fragments from Phillips Spring, in contrast, show no indication of any morphological changes beyond the parameters of the wild morphotype gourd profile. With a mean thickness value of 1.3 mm and a range of 0.82–2.0 mm, the 12 Phillips Spring *Cucurbita* rind fragments compare closely to wild *C. pepo* ssp. *ovifera* var. *ozarkana* (Figure 7), and the four peduncle scars from Phillips Spring provide a similar close match with the wild Ozark gourd (Figure 9). In addition, the four Phillips Spring peduncle scars have a wild morphotype circular outline and appear to be from small globular fruits (Figure 16). A wide range of changes in fruit morphology indicative of

Figure 16. *Cucurbita* fruit-end fragment from the Phillips Spring site, c 3000 cal. B.C. in age, which exhibits the small, circular peduncle scar typical of modern wild Ozark gourds (*C. pepo* ssp. *ovifera* var. *ozarkana*). Scale bar = 1 cm.

deliberate and sustained human selection do appear, however, in the 2,000 years that follow the c. 3000 cal. B.C. Phillips Spring seed assemblage and its evidence for initial domestication of *C. pepo* in the eastern woodlands. These changes include a diversification of fruit forms, surface lobing and warting, and an increase in both fruit size and rind thickness (Cowan, 1997:73). As outlined above, analysis of the Guilá Naquitz *Cucurbita* assemblage indicates that a similar temporal-developmental sequence framed the early history of the Mexican lineage of domesticated *C. pepo* (ssp. *pepo*). An initial automatic adaptive response to deliberate planting at 8000 cal B.C. (seed size increase) was followed over the next 2,000 years (8000–6000 cal B.C.) by changes in fruit morphology (size, shape, color) that appear to reflect deliberate human selection rather than the adaptive syndrome of domestication.

Unfortunately, it is impossible to document in much detail, in either region, the timing and sequence of these different changes in fruit morphology that are the result of directed and sustained human selection practices, because of the paucity of *Cucurbita* assemblages that fall into the periods in question (8000–6000 cal B.C. in Mexico, 3000–1000 cal. B.C. in eastern North America). As shown in Figure 13, the Guilá Naquitz assemblage does show a gradual and sustained increase in peduncle (and fruit) size over a span of 650 years, from 6450–5800 cal. B.C. But with this single exception, evidence is lacking for human selection of fruit form during these periods of early cultivation.

Even so, although fruit form changed dramatically in both Mexico and the eastern United States during the 2,000 years after initial domestication, clear differences between the two regions could well reflect different developmental contexts and levels of intensity of human manipulation and management of crop plants. In the eastern United States during the period from 3000 to 1000 cal B.C., some if not all early food-producing societies were living in resource-rich river valley settings in relatively substantial and sedentary communities and were experimenting with a wide range of indigenous species of seed plants, including four domesticates (squash, sunflower, marshelder, chenopod) and a number of others of economic importance (e.g., little barley, maygrass, erect knotweed) (Smith, 1992). In this broad context of increasing crop plant management, *C. pepo* underwent considerable evolution and diversification of fruit forms. Fruit not only substantially increased in size by 1000 cal. B.C., as reflected by a considerable increase in rind thickness, but also developed substantial diversity in shape. Prominent lobing also appeared, as did frequent surface warting.

In Oaxaca, in contrast, there is no evidence as yet for the existence of more permanent sedentary settlements in the 2,000 years after the initial domestication of *C. pepo* and no evidence of a food-production economy expanding to encompass other plants. *Cucurbita pepo* appears to have been the solitary domesticated food-crop plant grown as part of an otherwise stable and ongoing hunting-and-gathering economy (Flannery, 1986). In comparison to the eastern United States and *C. pepo* ssp. *ovifera*, the Mexican-lineage *C. pepo* ssp. *pepo* appears to have followed a less diversified and perhaps slower evolutionary pathway during its first 2,000 years as a domesticated crop plant. Rind thickness, for example, increased to a much lesser extent. Although peduncle (and fruit) size steadily increased, there was seemingly little diversity of fruit forms by 6000 B.C.—only a zucchini-shaped fleshy squash, with the admittedly small sample giving no hint of other shapes, surface warting, or lobing. Interestingly, the Guilá Naquitz assemblage indicates that relatively large peduncles and the pumpkin-orange skin color, both clear modern markers of *C. pepo* ssp. *pepo*, appeared early in the history of the lineage. In addition, the zucchini fruit form and fleshy fruit forms appeared early, whereas a proliferation of fruit shapes and a dramatic increase in seed size did not appear until much later in the developmental history of *pepo* squash in Mexico (Smith, 1997b)

CONCLUSIONS AND DISCUSSION

Richard MacNeish began his search for the origins of agriculture in Mexico more than 50 years ago. Over this half-century span, Thomas Whitaker and Hugh Cutler carefully and comprehensively documented the early history of cucurbits in the Americas through the analysis of archaeological plant remains. One of the most important and certainly the oldest such archaeobotanical assemblage analyzed by Cutler and Whitaker was the small collection resulting from Kent Flannery's excavation of Guilá Naquitz cave, which they considered late in their careers. Whitaker and Cutler first reported on the Guilá Naquitz cucurbits more than 25 years ago (Whitaker and Cutler, 1971) and then in greater detail a decade ago (Whitaker and Cutler, 1986) as part of Flannery's massive site report on Guilá Naquitz. In this 1986 volume, Whitaker and Cutler's brief study of the cucurbits from the cave contributed a critically important, central, and solitary element that supported Flannery's multiple-level interpretive framework for the initial transition from hunting-and-gathering to food-production economies in Oaxaca. Without Whitaker and Cutler's identification of domesticated squash (*C. pepo*) in the preceramic habitation zones of Guilá Naquitz, Flannery's entire explanatory structure would, in fact, be rendered moot. After all, it would be hard to sustain an unfolding drama about the shift from hunting and gathering to early food production in the absence of a domesticate to fill the leading role. As detailed in this reanalysis of the Guilá Naquitz cucurbit assemblage, Whitaker and Cutler's casting of domesticated *C. pepo* in a starring domesticate role in the cave's preceramic habitation zones is not only sustained intact, but reinforced and expanded.

Whitaker and Cutler's analysis of the Guilá Naquitz cucurbit assemblage was not comprehensive or exhaustive in scope, but after 50 years of analyzing archaeological cucurbits, it is perhaps not surprising that they tended to adopt, like Robert Dole in his last campaign, more of a "and so on and so forth, whatever" approach, rather than discussing in detail questions they had already addressed over the decades in many previous publications. The reanalysis presented here fills the logical gaps left open in their original analysis, both by providing a more detailed description of the Guilá Naquitz cucurbit materials and by resolving, through multiple lines of evidence, a number of essential yet unaddressed issues. In so doing, the present study provides a strong theoretical, methodological, and empirical foundation for the general conclusions reached by Whitaker and Cutler.

For an example of the relation of the current study to that of Whitaker and Cutler, consider the "elevation and environment" explanation they offered for the presence of *C. pepo* at Guilá Naquitz and the absence of *C. argyrosperma* and *C. moschata* (see earlier discussion), which is unaccompanied by any discussion of the criteria they employed to actually identify Guilá Naquitz specimens as *C. pepo* as opposed to either *C. argyrosperma* or *C. moschata*. Given that they published, more than 35 years ago, the first detailed key for distinguishing between archaeologically recovered seeds and peduncles of these three domesticated species of squash (Cutler and Whitaker, 1961), it is probably safe to assume that they employed, to their own satisfaction, the same general criteria in assigning the Guilá Naquitz peduncles and seeds to *C. pepo*. When the set of distinguishing characters developed by Cutler, Whitaker, and others (see earlier discussion) for taxonomic identification were updated and expanded for the reanalysis reported here, all the Guilá Naquitz specimens that could be assigned to the species level did in fact fall comfortably and unequivocally under the *C. pepo* label.

Similarly, although Whitaker and Cutler recognized the presence of both a wild *Cucurbita* gourd and a domesticated *C. pepo* squash in the Guilá Naquitz preceramic assemblage and provided specimen count information for both, they did not give any indication of the standards employed to distinguish between the two or give a detailed account of which seed and peduncle specimens were assigned to wild versus domesticated categories. In the current reanalysis, a set of general profiles for wild *Cucurbita* gourd morphotypes was developed to establish boundary conditions for separating wild and domesticated *Cucurbita* seeds and peduncles, and these conditions provided a clear empirical basis for identifying the initial appearance and subsequent development of domesticated *C. pepo* in the preceramic habitation zones of Guilá Naquitz. In so doing, the current study confirmed Whitaker and Cutler's key conclusion—that a domesticated *C. pepo* squash was present in the preceramic levels of the cave.

Whitaker and Cutler also identified a *Cucurbita* seed as providing the earliest evidence for domesticated *C. pepo* in the cave. Recovered from a Zone D context, the seed measured 10.0 mm in length and, based on a conventional ^{14}C age determination obtained from charcoal from the same excavation unit (F5), was assigned an age, by association, of c. 7,800 years ago (7840 ± 240 ^{14}C years

B.C.; Table 1). When compared with the seeds of Pleistocene and present-day wild *Cucurbita* gourds in the reanalysis reported here, however (Figure 15), the 10.0-mm-long Zone D seed fell well below the 12.0-mm ceiling for wild seeds and thus did not measure up to star billing as the earliest domesticated plant in the Americas. At the same time, however, two other seeds from Guilá Naquitz, 13.2 and 13.8 mm in length, clearly qualified for such a lead role. The AMS ^{14}C dates obtained directly for these seeds yielded dendrocalibration curve intercepts of 8025 and 7975 calibrated calendar years B.C.

In summary, it is reasonable to conclude that Whitaker and Cutler were not only essentially correct, but also remarkably accurate in their primarily and most important conclusion—that a domesticated *C. pepo* squash was present in the preceramic habitation zones of Guilá Naquitz Cave by c. 7800 B.C. In a broad interpretive context, Kent Flannery's detailed multilevel explanatory framework for the transition from hunting and gathering to initial food production in Oaxaca can be confirmed as firmly footed in a solid and well-documented empirical foundation. Flannery's acceptance of the research findings of Whitaker and Cutler was clearly justified. His sustained confidence in the age and undisturbed nature of the habitation deposits in this small and shallow cave was well placed. Clearly, then, this return to reconsider Guilá Naquitz more than three decades after it was excavated and more than a decade after its publication has been well rewarded and has underscored the wisdom of Flannery's "Paul Masson" approach—to "publish no site before its time." Complex, full bodied, demanding, yet rich with subtleties, Guilá Naquitz can be said to have aged very well, indeed.

NOTES

1. In the first few squares excavated in Guilá Naquitz Cave, the top three layers uncovered were labeled Zones A, B, and C. As the excavation was expanded, however, two additional layers intervened between Zones B and C in some areas of the cave. These intervening zones were labeled Zones B2 and B3, with Zone B labeled as Zone B1 in these excavation units. Thus Zone B in some squares equates with Zone B1 in others.
2. Robinson and Decker-Walters (1997:74) also noted that, at maturity, the white- and green-striped fruits of one species of wild gourd, *C. fraterna*, turns pale yellow with pale orange stripes, suggesting that orange coloration alone may not be a unequivocal marker of domestication (see Smith 1995a, 1995b:165, for an illustration of pale yellow and orange *C. fraterna* fruit).
3. The seven seeds not deposited in the INAH collections are those illustrated in Whitaker and Cutler (1986), suggesting that the specimens they photographed were not reunited with the rest of the Guilá Naquitz materials before their return to INAH. The present location of these seeds is unknown.
4. Deena Decker-Walters brought this to my attention and provided seeds of a modern Mexican cultivar of *C. pepo* from Veracruz, which had long, overlapping thread layers extending from ridge to ridge. These modern seeds in which the thread layer completely covered the seed margin closely resembled some of the Guilá Naquitz seeds as well as seeds of modern *C. moschata*.
5. The oldest AMS age determination on *Cucurbita* material from Phillips Spring was obtained on a 10.8-by-7.4 mm seed that yielded a date of 4440 ± 75 B.P. (3075 cal. B.C.) (ß 47293).

REFERENCES

Bozarth, S. R., 1987, Diagnostic Opal Phytoliths from Rinds of Selected *Cucurbita* Species. *American Antiquity* 52:607–615.

Burgess-Terrel, M., 1979, A Study of *Cucurbita* Material from Salmon Ruin, New Mexico. Master's thesis, Department of Anthropology, Eastern New Mexico University, Portales.

Cowan, C. W., 1997, Evolutionary Changes Associated with the Domestication of *Cucurbita pepo*: Evidence from Eastern Kentucky. In *People, Plants, and Landscapes,* edited by K. J. Gremillion, pp. 63–85. University of Alabama Press, Tuscaloosa.

Cowan, C. W., and Smith, B. D., 1993, New Perspectives on a Wild Gourd in Eastern North America. *Journal of Ethnobiology* 13:17–54.

Cutler, H., and Whitaker, T., 1956, *Cucurbita mixta* Pang: Its Classification and Relationships. *Bulletin of the Torrey Botanical Club* 83:253–260.

Cutler, H., and Whitaker, T., 1961, History and Distribution of the Cultivated Cucurbits in the Americas. *American Antiquity* 26:469–485

Decker, D., 1984, Squash Seeds from NAN Ranch, New Mexico. Unpublished manuscript, Archaeobiology Program, National Museum of Natural History, Smithsonian Institution, Washington, D.C.

Decker, D., 1986, A Biosystematic Study of *Cucurbita pepo*. Ph.D. dissertation, Department of Biology, Texas A & M University, College Station.

Decker, D., and Newsom, L., 1988, Numerical Analysis of Archaeological *Cucurbita pepo* Seeds from Hontoon Island, Florida. *Journal of Ethnobiology* 8:35–44.

Decker, D., and Wilson, H. D., 1986, Numerical Analysis of Seed Morphology in *Cucurbita pepo*. *Systematic Botany* 11:595–607.

Decker-Walters, D., Walters, T. W., Cowan, C. W., and Smith, B. D., 1993, Isozymic Characterization of Wild Populations of *Cucurbita pepo*. *Journal of Ethnobiology* 13:55–72.

de Wet, J. M. J., and Harlan, J. R., 1975, Weeds and Domesticates: Evolution in the Man-made Habitat. *Economic Botany* 29:99–107.

Doran, G., and Newsom, L., 1990, A 7,290-Year-Old Bottle Gourd from the Windover Site, Florida. *American Antiquity* 55:354–360.

Dumond, D. E., 1987, Review of *Guilá Naquitz: Archaic Foragers and Early Agriculture in Oaxaca, Mexico*, edited by K.V. Flannery. *American Anthropologist* 89:195.

Flannery, K. V., 1973, The Origins of Agriculture. *Annual Review of Anthropology* 2:271–310.

Flannery, K. V. (editor), 1986, *Guilá Naquitz, Archaic Foraging and Early Agriculture in Oaxaca, Mexico*. Academic Press, New York.

Fritz, G., 1994, Are the First American Farmers Getting Younger? *Current Anthropology* 35:305–309.

Harlan, J. R., and de Wet, J. M. J., 1965, Some Thoughts on Weeds. *Economic Botany* 18:16-24.

Harlan, J. R., de Wet, J. M. J., and Price, E. G., 1973, Comparative Evolution of Cereals. *Evolution* 27:311–325.

Heiser, C., 1985, Some Botanical Considerations of the Early Domesticated Plants North of Mexico. In *Prehistoric Food Production in North America*, edited by R. I. Ford, pp. 57–72. Anthropological Papers No. 75. Museum of Anthropology, University of Michigan, Ann Arbor.

Heiser, C., 1990, New Perspectives on the Origin and Evolution of New World Domesticated Plants: Summary. *Economic Botany* 44 (3):111–116.

Kaplan, L., 1993, Accelerator Mass Spectrometry Dates and the Antiquity of *Phaseolus* Cultivation. Paper presented at the Annual Meeting of the Society for Economic Botany, Miami, Florida.

Kaplan, L., 1995, *Phaseolus* Beans, Accelerator Dates in the Americas. Paper presented at the Annual Meeting of the Society for American Archaeology. Minneapolis, Minnesota.

King, F. B., 1985, Early Cultivated Cucurbits in Eastern North America. In *Prehistoric Food Pro-*

duction in North America, edited by R. I. Ford, pp. 73–79. Anthropological Papers No. 75. Museum of Anthropology, University of Michigan, Ann Arbor.

Long, A., Benz, B., Donahue, J., Hull, A., and Toolin, L., 1989, First Direct AMS Dates on Early Maize from Tehuacan, Mexico. *Radiocarbon* 31:1035–1040.

Merrick, L., and Bates, D. M., 1989, Classification and Nomenclature of *Cucurbita argyrosperma*. *Baileya* 23:94–102.

Nee, M., 1990, The Domestication of *Cucurbita* (Cucurbitaceae). *Economic Botany* 44(3):56–68.

Newsom, L., Webb, S. D., and Dunbar, J. S., 1993, History and Geographic Distribution of *Cucurbita pepo* Gourds in Florida. *Journal of Ethnobiology* 13:75–79.

Pickersgill, B., and Heifer, C., 1977, Origins and Distribution of Plants Domesticated in the New World Tropics. In *Origins of Agriculture*, edited by C. Reed, pp. 803–835. Mouton, The Hague.

Piperno, D., 1988, *Phytolith Analysis*. Academic Press, New York.

Roberts, K. M., 1995, *Cucurbita* ssp. and *Lagenaria siceraria*. In *Laboratory Guide to Archaeological Plant Remains from Eastern North America*. Department of Anthropology, Washington University, St. Louis.

Robinson, R. W., and Decker-Walters, D., 1997, *Cucurbits*. CAB International, New York.

Russell, P., 1924, Identification of the Commonly-Cultivated Species of *Cucurbita* by Means of Seed Characters. *Journal of the Washington Academy of Sciences* 14:265–269.

Smith, B. D., 1992, *Rivers of Change*. Smithsonian Institution Press, Washington, D.C.

Smith, B. D., 1995a, *The Emergence of Agriculture*. Scientific American Library, New York.

Smith, B. D., 1995b, The Origins of Agriculture in the Americas. *Evolutionary Anthropology* 3:174–184.

Smith, B. D., 1997a, The Initial Domestication of *Cucurbita pepo* in the Americas 10,000 Years Ago. *Science* 276:932–934.

Smith, B. D., 1997b, A Reconsideration of the Ocampo Caves and the Era of Incipient Cultivation in Mesoamerica. *Latin American Antiquity* 8:342–383.

Smith, B. D., 1998, *The Emergence of Agriculture*. W.H. Freeman, New York.

Smith, C. E., 1986, Preceramic Plant Remains from Guilá Naquitz. In *Guilá Naquitz: Archaic Foraging and Early Agriculture in Oaxaca, Mexico*, edited by K. V. Flannery, pp. 265–273. Academic Press, New York.

Spencer, C. S., and Flannery, K. V., 1986, Spatial Variation of Debris at Guilá Naquitz: A Descriptive Approach. In *Guilá Naquitz: Archaic Foraging and Early Agriculture in Oaxaca, Mexico*, edited by K. V. Flannery, pp. 331–368. Academic Press, New York.

Tapley, W. T., Enzie, W. D., and Van Eseltine, G. P., 1937, *The Vegetables of New York*, Volume 1, Part IV: *The Cucurbits*. New York Agricultural Experiment Station, Geneva.

Whitaker, T., 1981, Archaeological Cucurbits. *Economic Botany* 35:460–466.

Whitaker, T., and Bohn, G. W., 1950, The Taxonomy, Genetics, Production and Uses of the Cultivated Species of *Cucurbita*. *Economic Botany* 4:52–81.

Whitaker, T., and Cutler, H., 1971, Prehistoric Cucurbits from the Valley of Oaxaca. *Economic Botany* 25:123–127.

Whitaker, T., and Cutler, H., 1986, Cucurbits from Preceramic Levels at Guilá Naquitz. In *Guilá Naquitz: Archaic Foraging and Early Agriculture in Oaxaca, Mexico,* edited by K.V. Flannery, pp. 275–279. Academic Press, New York.

Wilson, H., 1989, Discordant Patterns of Allozyme and Morphological Variation in Mexican *Cucurbita*. *Systematic Botany* 14:612–623.

Chapter 3

The Evolution of the Tipití

A Study in the Process of Invention

ROBERT L. CARNEIRO

It is easy to be impressed by what one does not understand very well.
G. H. HARDY

INTODUCTION

Manioc is the staple crop of almost all Indian tribes in the Amazon Basin. However, many varieties of this plant are poisonous, containing lethal amounts of prussic acid in their tubers. To render the tubers fit for human consumption, most tribes of the region employ a type of press known as a *tipití*.[1]

The *tipití* is an extendable basketry tube made from reed strips, enclosed at the lower end and open at the upper. It is plaited on the bias, which is to say that the two sets of interwoven elements opposed to each other run diagonally in relation to the long axis of the tube. This type of plaiting allows a greater degree of stretching when the *tipití* is put to use.

In using this device to squeeze manioc, it is first compressed from both ends to widen the opening at the top. Through this opening, the tube is filled with grated manioc pulp. Once filled, the *tipití* is hung from the end of a beam by a loop woven into the upper end of the tube. Through a similar loop at the lower end, a pole is inserted to be used as a lever. One end of this lever is lashed to a vertical

ROBERT L. CARNEIRO • Department of Anthropology, American Museum of Natural History, New York, New York 10024.

Cultural Evolution: Contemporary Viewpoints, edited by Gary M. Feinman and Linda Manzanilla. Kluwer Academic/Plenum Publishers, 2000.

post, while the other is forced down by suspending a heavy weight from it or, more commonly, by having a person push down on the lever or sit on it (Figure 1).[2]

Pressing down on the lever stretches the *tipití,* thus narrowing its diameter. As it stretches, the inner volume of the *tipití* is reduced to about half of what it was when first filled with grated pulp.[3] The pressure to which the resulting lateral compression subjects the contents of the device forces the poisonous juice of the manioc pulp to be extruded through the interstices of the plaited mesh, dripping down the outside of the tube. After the pulp has been thoroughly squeezed, it is left compact, dry, and relatively free of prussic acid. Of all the devices used to detoxify manioc, the *tipití* is the most efficient and certainly the best known.

For many years, this device has attracted the attention and drawn the praise of anthropologists far and wide. More than a century ago, E. B. Tylor called it an "admirable discovery" (Tylor, 1871:1:57). Erland Nordenskiöld, a leading student of South American ethnology, described the *tipití* as "a very remarkable invention" (Nordenskiöld, 1929:9). William C. Farabee, another Amazonian specialist, called it "one of the most ingenious and most important inventions made by any primitive people" (Farabee, 1924:21).

More recently, in discussing the process of invention generally, Leslie White remarked:

> How the aborigines discovered that the removal of the poisonous prussic acid from the manioc root . . . could be done and how they perfected this technique is a matter of wonder. Ignorant of chemistry, knowing that initially the plant was deadly, and with minds full of magic, myth, and superstition, one wonders how they ever accomplished so difficult a feat. . . . We may, I think, regard it as one of the most difficult inventions in history (White, 1948:222n.).

Thus, so extraordinary has the *tipití* appeared to scholars that its invention has come to be regarded as not only remarkable but virtually unfathomable.

Without a doubt, the *tipití* is a very clever device. Nonetheless, as often happens when we are acquainted with only the final form of an invention and know nothing of its various stages of development, we find it difficult to imagine how it could have been devised. The tendency then is to attribute to its unknown inventors a measure of genius that may be out of all proportion to the degree of intelligence actually required. It is my contention that precisely this happened in the case of the *tipití.*

One of the premises that led me to undertake this chapter was a conviction that the invention of the *tipití,* like virtually all other inventions, was not the result of a single brilliant flash of intuition. Rather than the device having come into existence full blown, it seemed to me that it was the culmination of a series of steps in an orderly and intelligible progression. Indeed, I hold that not only is it unnecessary to invoke a stroke of genius to account for the *tipití,* but that its invention did not even require a superior degree of intelligence. The genius theory of invention, so often and so readily invoked, is far from being satisfactory, be-

Figure 1. The *tipití* shown in use by a Guiana Indian woman. Note manioc pulp protruding from the upper end and the poison-containing juice being caught in a small pot below (photograph by Borys Malkin).

cause accounting for an invention by attributing it to "genius" leaves the accomplishment essentially unexplained, "genius" being generally regarded as unaccountable and inscrutable.

As an example of the human tendency to attribute inventions to extraordinary intelligence, the American philosopher and psychologist, William James, speaking of the invention of the chronometer (among other "machines and instruments"), wrote: "Originally all these things . . . were *flashes of genius* in an individual head, of which the outer environment showed no sign" (James, 1917:196, emphasis added). In fact, the invention of the chronometer was no single "flash of genius" in the mind of its inventor, John Harrison, but the summation of 35 years of his painstaking work in seeking to perfect a timepiece that would be accurate to the degree stipulated by the British Admiralty in a competition that Harrison had entered (see Milham, 1947:261–268; Sobel, 1995).

In the case of the *tipití* we can, I think, make a convincing case that its development was a similarly gradual evolutionary process from very simple beginnings, one that proceeded by a series of small improvements until it attained its final fully evolved form. No major break in the sequence requires us to invoke a remarkable degree of insight to bridge it.

Following White's discussion of the invention of the *tipití*, just alluded to, he went on to observe: "Perhaps if we had a complete record of the discovery it might, and probably would, seem simpler" (White, 1948:222n.). What follows, serves, I think, to confirm White's acute surmise.

The Aim of the Chapter

This chapter aims to reconstruct the sequence of stages through which the *tipití* came into being, in as much detail as the evidence permits. The earlier stages of this evolution took place long before ethnographers were on the scene to record it. Moreover, the perishable nature of the materials used in manioc squeezers makes it unlikely that archaeological evidence bearing on its evolution will ever come to light. My reconstruction, therefore, must be hypothetical, based largely on inferences drawn from a study of the various surviving means by which manioc is detoxified. Nevertheless, although the relevant evidence is scattered and fragmentary and interpolation is required where such evidence is lacking, it still seems possible to piece together a coherent and plausible account of how the *tipití* evolved.

Detoxification of Other Food Plants

Before discussing the various ways in which the poisonous juice is removed from bitter manioc, it is important to note that the extraction of a toxic element from a potentially edible food is by no means uncommon among primitive peoples. Two examples suffice.

To free the wild *mungura* yam of its toxic substance, the Wanindiljaugwa aborigines of Groote Eylandt in northern Australia proceed as follows. After the root is shredded, it is enclosed in a loose container, which is then placed in a stream overnight, allowing water to run through it. The flowing water leaches out the toxin, permitting the nutritious pulp to be eaten the following morning (Worsley, 1962:164).

Acorns, as is well known, were a dietary mainstay of the Indians of central California. But acorns contain tannic acid, which must be removed before they can be rendered edible. The most common method of doing so was, as in the previous case, by leaching. After the shell of the acorn was cracked, the nut meat was reduced to a pulp. Then,

> When the meal was ground sufficiently fine, it was taken to the bank of a stream for the leaching process. Most frequently the meal was placed directly on the sand in a shallow depression or basin which had been prepared for the purpose. Then water was dipped from the stream and poured over and through the meal in the manner of making drip coffee. This leaching process was repeated until the bitter taste of the tannic acid was eliminated. (Driver, 1961:68–69; see also Heizer and Elsasser, 1980:91–93)

The occurrence of a poisonous or bitter substance in an otherwise nutritious food thus constituted a challenge, rather than a deterrent, to primitive peoples seeking to make it edible. What was true of primitive peoples, generally, also was true of those unknown Amazonian Indians who first sought to make a wholesome food from a poisonous root.[4]

EARLY FAILURES TO ACCOUNT FOR THE *TIPITÍ*

One reason the *tipití* is generally regarded as such a remarkable invention is that it is often thought to be the only device known to Amazonian Indians for detoxifying manioc. So knowledgeable an Americanist as Herbert Spinden once remarked: "A single technical process of extracting the poisonous juice of . . . manioc is found wherever the plant is cultivated" (Spinden, 1917:275). More recently, Wendell C. Bennett stated that "the process and equipment for preparing bitter manioc are similar wherever it is grown" (Bennett and Bird, 1949:43).

These statements, however, are erroneous. Although the *tipití* is indeed the most widespread, best-known, and most efficient device used for detoxifying bitter manioc, it is lacking among many tribes who also grow the plant. The latter use various other means to achieve the same end. The form, method of operation, and distribution of these other contrivances are not only of interest in and of themselves, but also afford invaluable clues as to how the *tipití* itself might have arisen.

To reconstruct the invention of this device, however, we must first consider certain related evidence. We need to know something of the physiology of the manioc plant, the way the tuber is treated before the extraction of its poison, and even something of the techniques of Tropical Forest basketry. Taking all these factors into account helps shed considerable light on the problem we are seeking to solve.

Given the *tipiti*'s importance to the native cultures of Amazonia, and the interest and wonder the device has engendered, it is surprising that not more has been done toward deciphering its history. In his monumental work, *An Introductory Study of the Arts, Crafts, and Customs of the Guiana Indians,* Walter E. Roth, the most meticulous student of Amazonian Indian technology, noted the existence of methods other than the *tipití* for processing manioc (Roth, 1924:217; 1929:30) and suggested, although rather casually and apparently without much conviction, two earlier stages in the development of this device. However, he readily admitted that "the scientific history . . . of its evolution is very unsatisfactory" (Roth, 1924:281). Nordenskiöld, another of the giants of South American ethnology, a man who probed the origin of many other items of Amazonian material culture, seems never to have tried to reconstruct the development of the *tipití.*

The first approach to a systematic survey of the various techniques employed by Amazonian Indians to remove the poison from bitter manioc was that of Alfred Métraux. Métraux noted the occurrence of manioc-processing devices other than the *tipití* among a number of Tropical Forest tribes, especially the inhabitants of the Upper Xingú,[5] but also among the Witoto, Yamamadí, and Kayapó. "This fact," he wrote, "indicates that the discovery of the . . . [*tipití*] does not date back to great antiquity" (Métraux, 1928:104). That is, so impressed was Métraux with the superiority of the *tipití* over other types of manioc squeezers that he thought the continued existence of these simpler forms among outlying tribes meant that not enough time had elapsed for the *tipití* to have supplanted them.

Plotting the distribution of the *tipití* on a map of Amazonia and noting that its use was particularly common in the area of the Guianas, Métraux surmised that it had been invented in that region, and that "from there, this invention seems to have spread along the Amazon and up its principal affluents" (Métraux, 1928:114–115, 104). Although Métraux drew attention to the existence of simpler forms of manioc squeezers, he did not attempt to trace the steps by which the *tipití* might have evolved from them. Neither did Lowie (1948:6), who nevertheless observed that "the basketry press [*tipití*] obviously presupposes earlier developmental stages."

More recently, Gertrude E. Dole (1960) undertook an extensive and detailed study of the distribution of manioc squeezers throughout Amazonia and drew historical conclusions therefrom. She also considered the problem of the evolution of the *tipití*, discussing how the various devices used for pressing manioc appeared to be related to its development. However, at the time of Dole's writing, she did not have available to her evidence that permitted a more complete and convincing genealogy of the *tipití.* The present study, while drawing freely on Dole's work, encompasses a somewhat larger body of materials than she covered and also deals with it in greater detail.

The original stimulus for reexamining the origin of the *tipití* came in 1958 with the collection by Anthony Leeds among the Yaruro of Venezuela of an unusual device for preparing manioc flour. To the best of my knowledge, this device, which the Yaruro call a *hanto*, has never been described or illustrated in print.[6] Although I no longer believe (as I once did) that the *hanto* represents a stage in the evolution of the *tipití*, it nevertheless seems fitting to mention it here as having provided the initial impetus for the present study.

THE PROBLEM OF DETOXIFYING MANIOC

To gain the necessary perspective on the development of the *tipití*, we must consider, if only briefly, the problems surrounding the original domestication and early use of manioc. What I noted above about the paucity of studies of the origin of the *tipití* also holds for studies seeking to discover the origin of cultivated manioc. In view of its importance as a food crop among peoples throughout the tropical world, it is surprising that so little has been written about its origin and early history. We are still unacquainted with manioc's wild ancestor (Sauer, 1950:507) and can only guess at the circumstances of its domestication.

Until fairly recently, it was believed that there were two species of manioc, generally called "bitter" and "sweet" and identified botanically as *Manihot utilissima* and *Manihot aipi* (or *palmata*), respectively. However, the work of the botanist David J. Rogers (1965), the leading student of the manioc plant, has shown that bitter and sweet manioc are merely different forms of the same species, which should properly be called *Manihot esculenta*. We still do not know, however, whether the tubers of the earliest cultivated manioc were poisonous, and if so, just how poisonous.

The toxic element in manioc is hydrocyanic acid (HCN), commonly known as prussic acid, a very powerful poison. In undisturbed manioc tubers, that is, in tubers still growing in the ground, prussic acid does not occur in free form, but only in the form of cyanogenic glucosides, principally linamarin. Only after the root is picked and an enzyme, linamarase, begins to break down the glucosides is prussic acid released (Cooke and Cock, 1988:65; Watt, 1908:767). This process is accelerated by grating, which, by causing "extensive disintegration of the plant tissues," leads to the enzyme and the glucosides coming in contact, thus "maximizing the conversion of bound to free cyanide" (Dufour, 1995:154).

In its pure liquid state, HCN boils at 150°C (Dufour, 1995:157) and is thus easily volatilized by the application of heat. Whether it is possible to eliminate all or substantially all the prussic acid in poisonous manioc tubers by roasting them whole or by boiling them, either whole or cut into chunks, is a question of some importance. If it is possible to do so, then the process of grating and squeezing out the poisonous juice from manioc roots might have been unnecessary to render them edible. Doing so might then have been for some other purpose, the most obvious one being to reduce the tuber to flour. Flour provided another form in which manioc could be eaten, as well as a convenient way in which it could be preserved by storage in a basket (see, e.g., Cook, 1903:489). Once picked and allowed to stand unprocessed, however, manioc tubers blacken and spoil in a few days. Thus manioc may at first have been grated and squeezed, not because it was the only way to detoxify it, but simply to prevent it from spoiling and to render it into the more useful form of flour.

However, if bitter manioc *cannot* be entirely freed of prussic acid merely by roasting or boiling, then grating the tuber and squeezing the pulp were probably first done to overcome its toxicity. In that case, whatever other advantages grating and squeezing conferred on the end product, they were secondary to the primary one of detoxification.

Whether manioc can be detoxified by roasting or boiling alone should be a relatively simple question to answer. Yet the investigations I know of bearing on this problem do not seem to be conclusive. One such study is cited by Y. R. Chadha, (1961:106): "Experiments in Tanganyika showed that, in one variety of bitter manioc, the hydrocyanic acid content of 100 g tubers was reduced from 39.6 to .7 mg after boiling, and in another variety of sweet manioc from 4.1 mg to .6 mg after roasting." These experiments support the belief that bitter manioc can be pretty well detoxified merely by boiling the tubers and perhaps also by roasting.[7] On the other hand, Dufour (1995:157), citing the results of other experiments, reported that "the simple methods of processing [manioc tubers] such as boiling and roasting are not very effective in reducing cyanide concentration."[8]

One of the few instances in Amazonia of the use of heat to drive off the prussic acid in whole ungrated manioc roots is reported by Kalervo Oberg (1953:22) for the Kamayurá:

> Another way of expelling the poisonous acid is by simply drying the tubers. The tubers are first peeled, but instead of being grated and pressed are dried near a fire. When thoroughly dried they can be stored in baskets or can be ground directly into meal in a mortar. By preparing manioc tubers in this way, of course, the by-products of juice and starch are not obtained.

However, as Oberg made clear, the Kamayurá use this method only as an infrequent alternative to the established method of grating and squeezing.

Few instances are known to me of Amazonian tribes who prepare bitter manioc tubers for eating by merely boiling or roasting. Moreover, Indians who raise both bitter and sweet manioc, as do the tribes of the Uaupés region, almost invariably process the two varieties differently (Koch-Grünberg, 1910:209). They grate and squeeze the bitter, but only roast or boil the sweet. In view of this near-perfect correlation between the two types of manioc and the two forms of processing, it appears that bitter manioc cannot with certainty be completely detoxified by boiling or roasting alone.

Whatever the incentive that gave rise to the practice, reducing manioc tubers to a mash or pulp was undoubtedly the first major advance in removing their prussic acid. At first, grating was probably done with the thorny aerial root of the cachapona palm (*Iriartea exorrhiza*), which is still used widely in Amazonia for this purpose.[9] The use of a palm-root grater for other purposes may well have preceded its application to manioc roots, because some tribes like the Amahuaca, who have only sweet manioc and do not grate it, use this thorny root to grate the wild *huito* fruit (*Genipa americana)*, whose juice yields a black dye.

Grating manioc tubers can be made easier, or even circumvented altogether, by first macerating the tubers in water. This process not only softens the tuber but also helps remove part of its prussic acid. Some of the acid dissolves in the water and is thus removed by leaching. Some of it, though, is dispelled by the partial fermentation that the tuber undergoes after soaking for several days (Dufour, 1995:155–156; Le Cointe, 1922:332–333). The Guaraní living around the Río de la Plata are said to have soaked poisonous manioc tubers in water to "lose the poison" and then reduced the softened pulp to flour (Vázquez de Espinosa, 1942:688).

What both grating and macerating accomplish is to break down the cellular structure of the manioc tuber, allowing the starch and the poison-bearing juice to be readily separated from the tissues of the root. By whatever means this is carried out, once bitter manioc has been reduced to a mash or pulp, it is ready for the next step in making it edible.

If there was ever a time when Amazonian Indians boiled manioc pulp immediately after grating to volatilize the prussic acid directly from the mash without first squeezing it, little evidence of this stage survives. In fact, I know of only one report of this practice. When the Umotina wish to make beer (instead of flour) from manioc, they boil the pulp immediately after grating it (Oberg, 1953:107). Among the vast majority of Tropical Forest tribes, though, the grating of manioc is followed directly by some form of squeezing to separate the poison-bearing juice from the coarse flour. It is with the first attempts to squeeze the juice from manioc pulp that our reconstruction of the invention of the *tipití* must begin.

EARLIER FORMS OF SQUEEZING MANIOC

The simplest and most obvious method of squeezing grated manioc is with the hands, and it seems likely that this was the means first employed for doing so. Nor is this altogether a surmise, because manioc pulp is still squeezed in the hands by a few tribes. The Nambikuara (Oberg, 1953:91), the Karajá (Ehrenreich, 1948:41; Lipkind, 1949:181), the Kajabí (Carneiro, 1953), and the Tapirapé (Wagley, 1977:58), among others, have done so in recent times.

Regardless of the way it is reduced to a mash, whenever manioc pulp undergoes squeezing, it separates into two components. One of them is a mass of coarse, more-or-less dry flour, and the other is a watery residue of juice. This juice contains not only the prussic acid, but also the fine starch flour known as tapioca, which is held in suspension. The earliest peoples to grate and squeeze bitter manioc might well have been content to use only the coarse flour. They may have thrown away the poisonous juice containing the tapioca, just as the nonpottery-using Mentuktire ("Chukahamay"), a Kayapó subgroup, did in the 1950s (Cláudio Villas Boas, personal communication, 1953).

But some attempt to recover the tapioca from the watery juice may have been made from the very beginning. Even the primitive Nambikuara, who, like the Mentuktire, lack pottery, pour the juice squeezed from the grated manioc mash "into a slight depression in the ground, and after the liquid has drained off . . . collect the thin film of starch flour" (Oberg, 1953:91). The fine tapioca flour can also be separated from the juice by pouring the juice into a gourd or calabash and allowing the tapioca to settle to the bottom, after which the poisonous liquid can be carefully decanted. What little prussic acid remains in the fine flour can probably be driven off by drying in the sun and certainly by baking on a griddle. Thus it is conceivable (although perhaps not probable) that the use of both fine and coarse flour goes back to the earliest days of manioc processing.

How the poisonous juice is processed once it has been expressed from the grated pulp is in itself an interesting facet of the problem. However, because it comes *after* the squeezing of the grated pulp, it lies outside the scope of this chapter.

Although squeezing grated manioc with the hands is a simple way to press out the juice, it has its limitations, the principal one being that only small amounts of mash can be squeezed at a time. For this reason, the first improvement in dealing with manioc pulp was probably aimed at compressing a larger amount of it at a time. One way in which this could have been done was by wrapping the pulp in a large leaf and squeezing the contents with both hands.

The idea of using a leaf to squeeze poisonous manioc probably occurred readily. In preceramic times, good-sized leaves were no doubt used as temporary receptacles for manioc pulp, either before or after squeezing. In picking up a lump of manioc mash with a leaf, at least a small amount of pressure would inevitably be applied to the contents, and some liquid would be extruded from it thereby. This would surely have been noticed and would have suggested the use of a leaf instead of just the hands for squeezing the pulp.

According to Cláudio Villas Boas (personal communication, 1953), the Mentuktire, when first contacted by him in 1953, squeezed manioc pulp in a banana leaf. The banana is generally regarded as a post-Columbian introduction into Amazonia, but in pre-Columbian times some other large leaf would probably have been used for this purpose. The most likely choice would have been the leaves of various species of *Heliconia*. Although somewhat smaller than banana leaves, the leaves of this plant are stronger and are still widely used throughout the region for lining baskets and for wrapping and handling food, as they were when Europeans first arrived (Oviedo, 1851:276–277).

But leaves are delicate and fragile. Thus a distinct improvement over a leaf for squeezing manioc pulp would have been to use a sheet of bark cloth. This sheet would have been handled in the same way as a leaf, namely, by laying the grated pulp on top of it, rolling it up, and squeezing with the hands. Being stronger than a leaf, bark cloth would allow more pressure to be applied to the contents, another distinct advantage. Moreover, a sheet of bark cloth could be made to a size and shape more convenient for the purpose than that of a leaf.

Whether the use of a bark cloth sheet for squeezing manioc was indeed an early stage in the development of the *tipití* is debatable, perhaps even improbable. Only one ethnographic instance of the use of bark cloth to squeeze manioc seems to be recorded for South America. Furthermore, bark cloth has a limited and sporadic distribution in Amazonia. Its use for squeezing manioc, where this occurred, may thus be only a side branch in the evolution of the *tipití* rather than on the main trunk of its development.

A few Tropical Forest tribes squeeze manioc pulp with a piece of woven cotton cloth. I myself have seen this done by the Kuikuru of central Brazil, who make use of this method only as an occasional variant to their normal way of squeezing manioc with a mat strainer. This use of cotton cloth is also reported for the Bakairí (Oberg, 1953:72), Nambikuara (Oberg, 1953:91), and Umotina (Oberg, 1953:107).

The use of cotton cloth to squeeze manioc pulp has the same advantages over a leaf that bark cloth does. Still, it seems unlikely to me that this manner of squeezing manioc can be considered a stage on the direct line leading to the *tipití*. What argues against it is that woven cotton cloth probably appeared in Amazonia well after the introduction of bitter manioc. In fact, the weaving of cotton into cloth even today is absent from a number of tribes that grow bitter manioc. Moreover, the tribes cited above as using cotton cloth for squeezing manioc apparently do so with cloth of Western manufacture. The practice, then, may be a recent one.

The next stage in the development of manioc squeezing may well have been wrapping a strip of bark spirally around a mass of grated pulp and wringing the juice out of the mash by twisting the ends of the strip in opposite directions. This technique is used today by a few tribes, including the Nambikuara (Lévi-Strauss, 1948:363) and the Umotina (Oberg, 1953:107).

The spiral strip method is superior to the use of a leaf in squeezing grated pulp in that, being much stronger, a bark strip permits one to apply considerably greater force. Moreover, by wrapping such a strip around the mash and twisting it, one can transmit pressure evenly to all the pulp enclosed in the coils, instead of merely to that portion lying directly beneath the hands. Thus the use of a bark strip would have enabled more pulp to be squeezed at one time. In addition to involving the use of a new material—a bark strip replacing a leaf—there is another distinction to be noted here. Mechanically, a new method of squeezing was being employed, one that used a different physical principle—torsion—instead of the simpler one of applying direct downward or inward pressure with the hands.

It is not immediately apparent how the wrapped bark strip developed from the use of a leaf. I may, however, offer a conjecture. No doubt there always would have been some tendency to *twist* the lump of manioc held in the leaf as it was being compressed in the hands, just as we do in wringing out a washcloth. The realization that twisting was a more effective way of applying force than merely squeezing, along with the tendency for a leaf, or even a sheet of bark cloth, to tear (or come unwrapped), may have led to the innovation of using a spirally wrapped bark strip in applying pressure to the grated pulp. Bark strips were no doubt already widely used for lashing poles together, and their use in this new way would have been readily apparent.

Another type of squeezer employing the principle of torsion is found in the Guianas, where this technique is not used in squeezing manioc. Here, the bitter seeds of the *mora* tree, after being grated, are squeezed inside a bunch of palm-leaf midribs that have been wrapped around the pulp of this fruit and the ends twisted in opposite directions (Roth, 1924:217-218).

Since the above lines were first written, an artifact has come to light in the storerooms of the American Museum of Natural History showing that this same technique was actually employed to squeeze manioc by at least one tribe. The little-known Shimaku of northern Peru used, for this purpose, a simple device consisting of several dozen palm-leaf midribs stripped of their leaflets and gathered in two parallel bunches. Some four of the midribs were interwoven loosely through the rest to keep each bunch of midribs more or less together and at the same time separated from the opposite one. One of the long braided tails that formed the ends of this device was tied to a house post, after which the two sets of midribs were separated, and grated manioc pulp placed in the trough thus formed. The other braided end was tied around a stout stick that was then twisted, forcing out the poisonous juice (Figure 2) (Tessmann, 1930:496-497).

Figure 2. Shimaku manioc squeezer made from palm-leaf midribs (AMNH specimen 40.0/3724).

The Shimaku manioc squeezer is a very crude and no doubt inefficient device, probably allowing a good deal of the coarse flour to dribble out between the midribs. However, it is of very great interest in representing an early stage in the use of torsion to squeeze manioc pulp. Although this device may not have survived elsewhere in Amazonia as a manioc squeezer, it may nevertheless have once been widespread and led directly to the next stage.

THE KAYAPÓ TORSION SACK

A more advanced device for squeezing manioc, one that likewise employs the principle of torsion, must now be considered because it probably represents the next step in the development of the *tipití*. This device, which I call the Kayapó torsion sack, has a restricted distribution, being confined almost entirely to the Ge-speaking tribes of central Brazil. These tribes include the Apinayé, Sherente, and Eastern Timbira (Lowie, 1946:481; Nimuendajú, 1939:88), as well as various Kayapó subgroups such as the Xikrín (Frikel, 1968:40, plate 42; Krause, 1911:388; Ribeiro, 1988:57).[10]

To make the Kayapó torsion sack, the leaves of a *burití (Mauritia)* palm, with the midribs stripped off, are first plaited to create a flat web some 50 to 70 cm (20 to 28 in) long and about 13 to 25 cm (5 to 10 in) wide. This web is twill plaited on the bias so that it stretches more readily when twisted. The ends of the web are then gathered up and wrapped around with cotton cord, thus forming two straight, ropelike handles 10 to 12 cm (4 to 5 in) long, which afford a grip when using the device (Figure 3) (Frikel, 1968:40).

Although woven as a flat web, during the course of making the device, when the ends of the Kayapó torsion sack are gathered up and wrapped with cord, it assumes a concave, troughlike shape. Into this concavity, grated manioc is placed. With both hands, the two handles are then twisted in opposite directions, causing the sack to coil tightly around the manioc mash (Krause, 1911:388). The twisting force applied compresses the contents and forces the poison-containing juice out through the interstices.

The Kayapó torsion sack is short enough to be used by one person, as shown in Figure 4, which illustrates a specimen of this device extended (above) and then twisted spirally as if in use (below). Although the amount of manioc mash it can squeeze at one time is not large compared with the capacity of the *tipití*, it is greater than the quantity one can wring out with the hands alone or with a leaf, and probably greater than the amount that can be squeezed by the Nambikuara's bark strip or the Shimaku's midrib squeezer.

In some ways, the Kayapó torsion sack may be said to be a forerunner of the *tipití*. It includes three elements that at some point in its development were incorporated into the *tipití* itself: (1) the use of a plaited mesh with interstices through

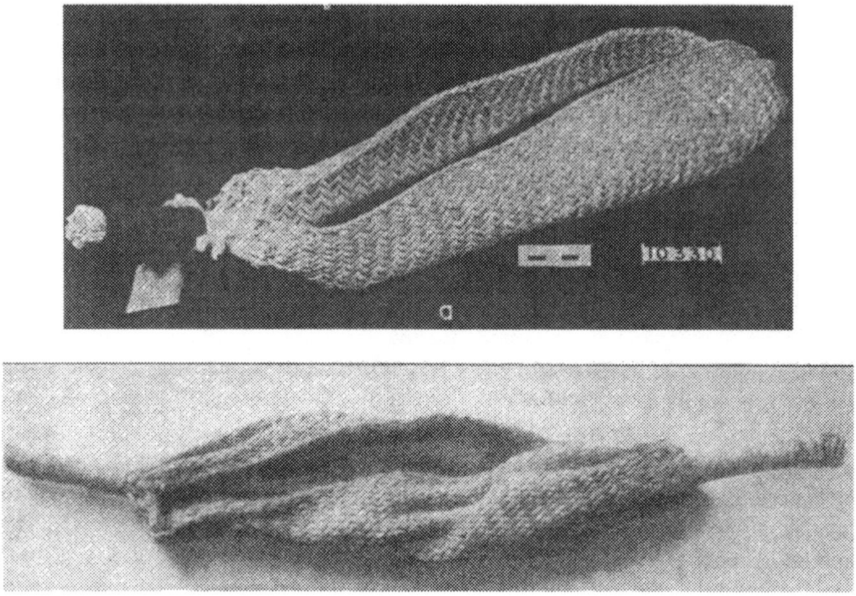

Figure 3. Kayapó torsion sack. Above, a specimen from the Xikrín (Frikel, 1968:40); below, a specimen from an unidentified Kayapó group (Krause, 1911:facing p. 392).

which the poisonous juice can be extruded; (2) the plaiting of this mesh on the bias, giving it greater flexibility; and (3) the forming, during use, of something approaching an enclosed cylinder.

THE KOFÁN AND SIONA SQUEEZERS, AN ADVANCE

Except for the Ge-speaking tribes of central Brazil, the torsion sack is all but unknown in Amazonia. However, we find among the Kofán of southern Colombia and the Siona of adjacent parts of eastern Ecuador a device that appears to represent a step beyond the Kayapó torsion squeezer in the line of development leading to the *tipití* (Figure 5).

This device can be thought of as basically a flat web whose ends have been gathered up and wrapped to form two straight handles, just as with the Kayapó sack. Like the latter, after manioc pulp is placed inside, the handles are twisted in opposite directions. There are, however, two noteworthy differences in the construction of the two devices. For one thing, the torsion sack of the Kofán and Siona is plaited from bast (inner bark) strips instead of palm leaves. More important, though, it is not plaited as a flat web along its entire length, but is woven in

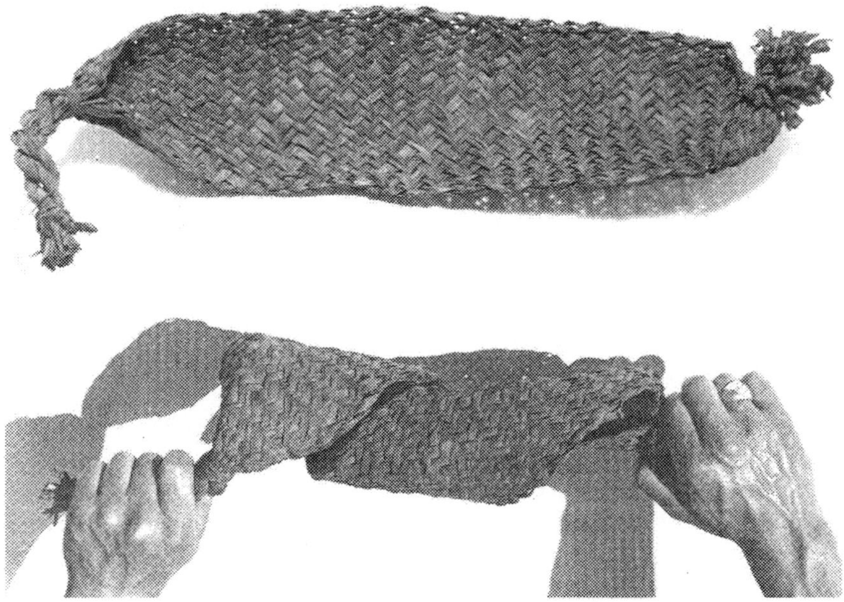

Figure 4. Torsion sack of the Kayapó type once in use among the Sherente. Shown extended (above) and spiraled as if in use (below).

such a way as to form tubular pockets immediately inside the two handles, a long slit being left open along the top through which it is filled. This device thus shows the first steps toward an intentionally woven cylinder and in this regard may be said to foreshadow the *tipití*.

THE "PUTTEE" FORM OF MANIOC SQUEEZER

Although torsion represents a significant advance over simple squeezing in dealing with manioc pulp, we are still a long way at this stage from the fully devel-

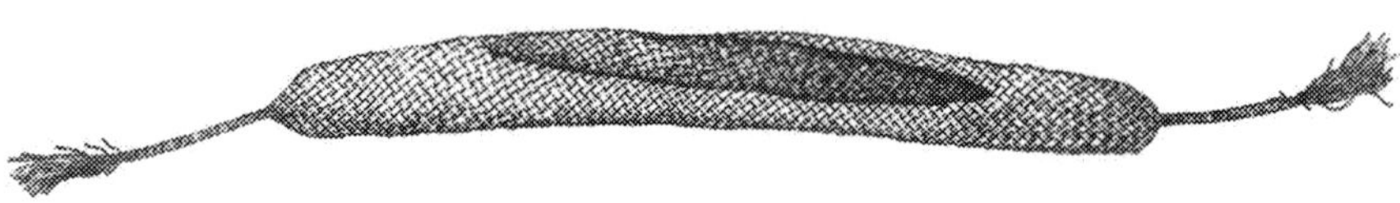

Figure 5. Kofán torsion sack, plaited from bast strips, with pockets at each end and open along the top (AMNH specimen 40.1/3913).

oped *tipití*. Nevertheless, we need to take a closer look at torsion-based devices and see how they may have led to further advances in manioc squeezing.

To the best of my knowledge, the first written account of the Kofán/Siona device described above was given by W. E. Hardenburg in *The Putumayo, The Devil's Paradise,* published in 1912. Discussing the "Cioni" (Siona) method of manioc processing, Hardenburg wrote:

> In preparing the *fariña*, the *yuca* is thrown into a trough filled with water and left there until it is in a state of semi-putrefaction, when it is taken out, peeled, and pulverised. . . . If it is the *yuca brava*, the poisonous sap must first be removed. To do this, the *yuca*, already pulverised, is wrapped up in a good-sized piece of *llanchama*—the tough, inner bark of a tree of the same name—which is then twisted up and tightened with a stick, after the fashion of a tourniquet, until the sap is all pressed out. (Hardenburg, 1912:84)

Here an interpolation seems in order. The word *llanchama* is widely used in eastern Ecuador and Peru to refer to bark cloth. Thus, the immediate interpretation of Hardenburg's passage seems to be that the Siona squeezed manioc pulp in an unwoven sheet of this material, but another interpretation is possible and, it seems to me, more likely. The word *llanchama*, applied in this region to bark cloth, comes originally from the name of the tree (*Sterculia* sp.?) from whose inner bark it is commonly made, but in addition to providing the raw material for bark cloth, the bast of this tree is put to other uses. For example, strips of it are used for lashings and other purposes. Hardenburg may have used the word *llanchama* loosely—for a web plaited of bark strips obtained from this tree rather than for a plain sheet of bark cloth as such. His mention of the use of sticks in working this device makes this interpretation even more likely.[11]

Nothing crucial hinges on it, however. If it *was* a sheet of bark cloth and not a plaited web that the Siona were using to squeeze their manioc, that would merely constitute an example of a device that we have already posited as an earlier evolutionary precursor of the *tipití*.

Three years after Hardenburg, Thomas Whiffen published a description of what is probably the same type of manioc press, which he observed being used by the Witoto during his visit to that tribe in 1908 or 1909. Whiffen gave two separate but similar descriptions of the Witoto squeezer:

> Instead of this cylinder [that is, the *tipití*] the Witoto use a long web, a rectangular strip about ten inches wide of plaited bark-fibre, about an inch wide [referring to the width of the individual strips]. This they wind round the grated manioc after the manner that puttees are adjusted on the leg. The tighter they twist the pliable web the greater the pressure upon the crushed roots, and the juice is thus wrung out of them. (1915:98–99)[12]

> [In squeezing grated manioc pulp] The Witoto and other tribes on the south [of Colombia] use a long rectangular palm-fibre [probably bast] wringer,

> which is twisted to form a cylinder in the same way as a puttee is wound round the leg. In this elastic cylinder it is compressed till all the poisonous juice has been drained away, when the remainder, a coarse kind of flour, is placed in an open pan and left to get thoroughly dry. (1915:132)

Whiffen said nothing about the insertion of sticks through loops at either end of the "puttee" (as I call this device) to assist in twisting it. However, in Tessmann's (1929:319–320) brief description of Witoto manioc processing, which he observed around 1928, he mentioned the use of such sticks. In addition, the photograph of a puttee in Tessmann's book shows the remains of a crudely made loop at one end of the device into which a stick was most likely inserted (Figure 6). Moreover, in a photograph of a used Witoto puttee, shown in Whiffen (1915:facing 134), the braided ends appear shredded, or at least abraided, as if they had once formed some sort of loop through which a stick had been inserted and twisted with considerable force (Figure 7).

Neither Hardenburg nor Whiffen described just how the puttee was held while being twisted. We can infer, however, that it was held horizontally, about chest height, and that it could be used by either one or two persons. If the puttee was short enough, one person alone could hold a stick in each hand and twist them in opposite directions, producing the tourniquet effect described by Hardenburg. If the puttee was longer, it would probably have been used by two persons working together, again twisting their sticks in opposite directions.

It is important to note here that in twisting the ends of the puttee with the aid of a pair of sticks instead of just the hands, these Northwest Amazon tribes had

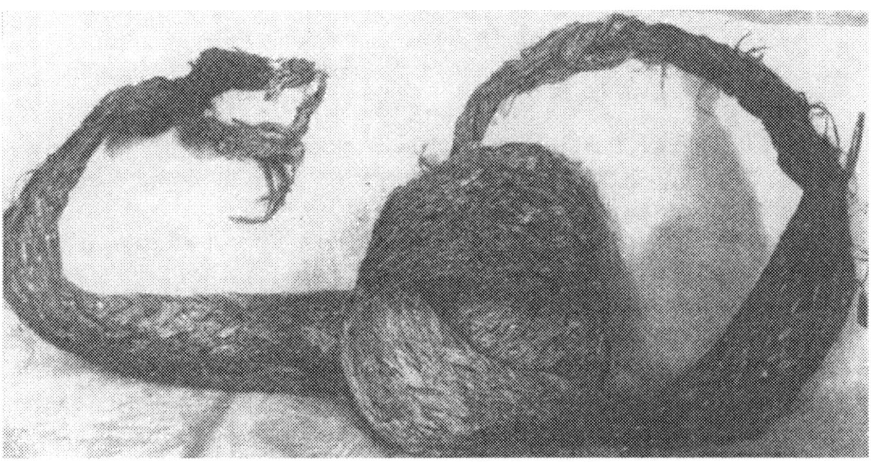

Figure 6. Witoto puttee shown after having been used. Note abrading of the end at the left, indicating twisting of the device with a stick (Tessman, 1930:facing p. 320).

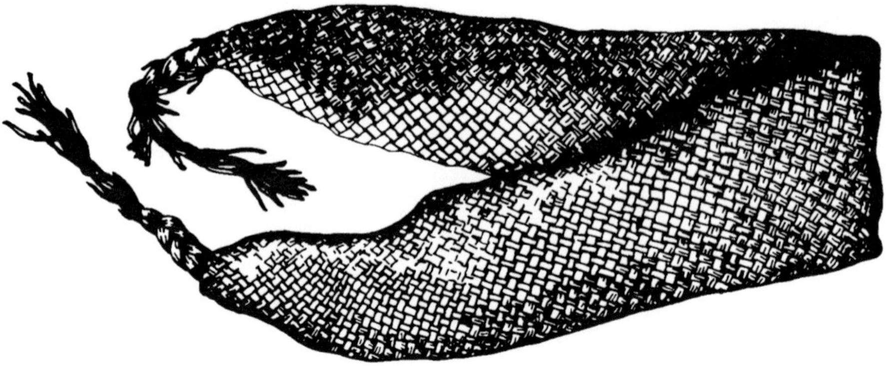

Figure 7. The Witoto puttee with abraded ends indicating twisting of the device with a stick (Whiffen, 1915:facing p. 134).

taken a significant step beyond the Kayapó torsion sack. The use of sticks no doubt allowed the application of a great deal more force in squeezing manioc pulp than the hands alone could provide and thus assured a more thorough detoxification of the grated mash.

That the type of squeezer just described may once have had a considerably wider distribution is suggested by the fact that the Yamamadí, living on the upper Purús some 950 km to the southeast, were said by Paul Ehrenreich (1948:104) in the 19th century to use a manioc squeezer whose description makes it sound much like that of the Witoto:

> Of singular manufacture . . . is the device used to press manioc *(kainaro)*. It consists of a concave, plaited strip of bast fiber 1 1/2 meters [5 ft] in length by 8 cm [3 1/8 in] in width, whose extremities end in narrow strips. To squeeze manioc, the pulp is placed on the concave inner surface, and the web is twisted spirally.

However, Ehrenreich failed to make clear whether the Yamamadí device was squeezed entirely by hand torsion or whether sticks were employed to provide a greater twisting force.

The puttee is similar to the Kayapó sack in that it works by torsion, but there are clear differences as well. The puttee is larger and is plaited from bast strips instead of palm leaves. Furthermore, it is a long web wrapped spirally around the manioc pulp rather than a concave receptacle into which the pulp is placed before squeezing. All things considered, in the evolution of manioc-squeezing devices, the puttee used by the Witoto seems to represent a step or two beyond the Kayapó torsion sack and has in fact four advantages over it:

1. It is made of stronger material.
2. It can squeeze more manioc pulp at one time. In fact, in an experiment carried out by Desidério Aytai with specimens in the collections of the

Museu Paulista, he found that the load of manioc pulp that could be put into a Witoto puttee was 8.2 liters, whereas the average load of two Kayapó torsion sacks he measured was only 0.7 liters. Thus, per load, the puttee can squeeze almost 12 times more manioc than the torsion sack (Aytai, 1987:120–121).

3. Because of the tourniquet effect produced by the wrapping of a web around the manioc mash and the use of sticks to tighten it, more pressure can be exerted on the enclosed mass and thus more poisonous juice can be squeezed out.
4. When twisted by two persons instead of one, considerably greater pressure can be exerted than is possible with the Kayapó torsion sack, which is twisted by only one person.

We can also say that, being more nearly tubular in form when it is coiled in use, the puttee is one step closer to becoming a *tipití.*

FURTHER RESEARCH ON THE PUTTEE

The area of Amazonia where the Witoto, Kofán, and Siona are located seemed to me so critical in unraveling the history of the *tipití* that when, in 1961, I learned that Borys Malkin was planning an ethnographic expedition to this region, I urged him to observe and record the exact manner in which the Witoto processed bitter manioc. During this expedition, and subsequent ones undertaken in 1966 and 1970, Malkin was able to do so among both the Witoto and the neighboring Kofán. The descriptions that follow are based on his detailed and meticulous observations, described in four brief publications (Malkin 1975a, 1975b, 1975c, and 1975d) and on specimens he collected.

In making their puttee, the Witoto first cut long strips of bast from a tree that may well be *llanchama*. A dozen or so of these strips, each about 2 cm (3/4 in) wide, are then plaited together on the bias to form a web some 30 to 38 cm (12–15 in) wide and 3 to 3.5 m (10–12 ft) long.[13] At the ends, the unplaited strips are gathered together and braided into a sort of "tail."

Once finished and ready for use, the puttee is extended, full length, on the floor of the house or on the ground outside, and covered along its entire length with grated manioc pulp. Beginning at one end, the puttee is then rolled up spirally until it forms a tube. At this point, the edges of the web touch each other but do not overlap. Holding the coiled-up tube carefully so that it does not unravel, two women carry it to a part of the house with a projecting beam. The braided tail at the upper end of the puttee, which has been tied into a loop by folding it on itself and then wrapping it a few times with a piece of cord, is engaged around the end of the beam. The tail at the lower end of the puttee is also twisted into a loop into which a long stick is slipped. The puttee is then pulled out diagonally from the house, and the stick inserted through the lower loop is twisted, the pressure thus generated squeezing out the poisonous juice (Figure 8).[14] A vessel may be placed below the squeezer to catch the extruded juice as it trickles down the sides of the device.

After it has been in use for a while, the Witoto puttee becomes somewhat longer, narrower, and distinctly concave. (Aytai [1987] offered an ingenious and convincing explanation of the physical reasons that an initially flat web should become concave with continued use, but it is unncessary for our purposes to go into it here.)

The manioc squeezer used by the Kofán is basically similar to that of the Witoto, but with some suggestive differences. In making this device, strips of bast 3 cm wide are removed from a certain tree (probably *llanchama*), placed on the ground, and then plaited together into a web. As among the Witoto, this web is plaited on the bias, both sets of strips being interwoven diagonally relative to the long axis of the web (Figure 9).

Compared with that of the Witoto, the Kofán puttee is relatively short, usually only 1.5 to 2 m long. Although for most of its length the Kofán squeezer is woven as a flat web, toward each end the bast strips are plaited to form a cylindri-

Figure 8. Witoto puttee in use. Upper end is attached to a special framework rather than being hung from the end of a house beam, as is usually done. Note the use of a long stick as a lever and a post as a fulcrum in operating the device (photograph by Borys Malkin).

cal pocket. The effect of plaiting these pockets is to make the rest of the web assume a concave shape, like a shallow trough (Figure 9).

Not all the bast strips are used to close up the ends of the Kofán puttee, however. Some are partially shredded, gathered together, and wrapped to form a permanent loop at opposite ends of the squeezer. These are distinct loops, not just tails, tied up to form a loop, as among the Witoto.

When the device is to be used, two women stretch it out on the ground, and one woman fills it with grated manioc pulp, patting and smoothing down the pulp to fill the device as full as possible. The women then carefully coil the ends of the puttee in opposite directions, thus wrapping the web around the enclosed mash. Then, inserting a stout stick into the loops at either end and twisting them in opposite directions, the women squeeze the pulp within. As they do so, the poison-containing juice is forced out through the interstices of the mesh. Sometimes the women relax the pressure for a while and then resume twisting. This proce-

Figure 9. Kofán man plaiting a short manioc squeezer of the puttee type but with pockets at each end as well as specially made loops for the insertion of sticks to be used in twisting the device (photograph by Borys Malkin).

dure may be repeated several times until as much of the juice has been extruded as possible (Malkin, 1975b:9-10) (Figure 10).

If she is working alone, a Kofán woman loops one extremity of the squeezer over the end of a protruding house beam. Then, with a stout stick inserted through the lower loop, she twists the device tightly, as before.

Whichever way the Kofán puttee is used, the poisonous juice extruded through the mesh is allowed to fall to the ground. Unlike most tribes, the Kofán apparently make no effort to collect the poisonous juice for further processing.[15]

FROM THE PUTTEE TO THE *TIPITÍ*

The Witoto, Kofán, and Siona puttees have much to tell us in our quest to trace the origin of the *tipití*. When the web that makes up the puttee is wrapped around grated manioc pulp, it assumes a cylindrical form, the shape of the classic *tipití*. Furthermore, as we have seen, it is the narrowing of this cylinder when the ends of it are twisted that causes it to constrict the enclosed mash, just as in the case of the *tipití*. However, there is an important difference here. With the puttee, this narrowing is produced by *torsion*, whereas in the *tipití* it results from *extension* caused by applying a downward force by means of a weight or a lever. Moreover, twisting the puttee causes it to *shorten*, whereas applying a downward force to the lower loop of a *tipití* causes it to *lengthen*.

Another difference between the two devices is that the puttee, when used, is often held horizontally, whereas the *tipití* squeezes its contents while suspended vertically. When the puttee is hung vertically—as is now often done by both the Kofán and Witoto, but was not reported by either Whiffen or Hardenburg—a further step in the progression toward the *tipití* has been taken.

The only major gap remaining to be bridged between the puttee and the *tipití* is that of replacing an initially flat web, subsequently coiled to form a cylinder, with a device that, from the outset, is plaited as a cylinder along its entire length. Because the manioc squeezer he or she was using assumed the form of a cylinder when being twisted, it must have occurred to some thoughtful individual to *begin* with a cylinder. As we have seen, a step in this direction had already been taken by the Kofán (and Siona as well), who in making their puttee plaited pockets at either end. The effect of these pockets is to give the device a trough-like shape, bringing it that much closer to the closed cylinder of the *tipití*. A progressive extension of the closure of the cylinder, from each end toward the middle, would eventually have led to a fully closed tube, thus, in essence, arriving at the *tipití*.

As already noted, the Kofán also plait loops at the ends of their puttee instead of merely braiding straight tails, as the Witoto do (Figure 9). This feature, that is, having distinct loops at either end of the squeezer, is of course a character-

Figure 10. Two Kofán women twisting a puttee using short sticks introduced through the loops at each end (photograph by Borys Malkin).

istic of the full-fledged *tipití*, and thus, in this regard as well, the Kofán squeezer has inched a bit closer to the final form of the *tipití*.

The main mechanical change involved in converting the puttee into the *tipití* was replacing torsion with stretching as the means by which pressure was exerted on the enclosed pulp. Although in terms of physical principles this was a major difference, it may nonetheless have been a relatively easy step to take.

We have seen that the flat web of the Witoto and Kofán puttee was already plaited on the bias, thus allowing a certain amount of stretch. One would expect, therefore, that the first tubular *tipití* would have been plaited in the same way. This method of plaiting, which had already proved advantageous when employed in the puttee, was even more important to the workings of the *tipití*. Only when plaited on the bias does the cylindrical barrel of the *tipití* stretch and narrow enough for efficient squeezing of the contents to occur when a downward force is applied to it.

More important, perhaps, was the fact that suspending a puttee vertically (or at least diagonally) was a technique already used by tribes like the Kofán and Witoto to allow one person, working alone, to operate the device. *Hanging* a manioc squeezer from a projecting beam would thus have been in the minds of those unknown Indians whose experiments led eventually to the *tipití*.

Initially, the downward force needed to operate the *tipití* by stretching it might have been provided by simply hanging a heavy weight from its lower end, as some tribes still do, but another and more efficient means of stretching the cylinder was soon hit on. This involved introducing a long pole through the lower loop of the *tipití*, attaching it at one end to a house post, and using it as a lever. Because inserting a stick through one or both loops was already a feature of the puttee, doing so with the *tipití* would have readily come to mind. Given the mechanical advantage made possible by the use of a lever, the force applied in squeezing manioc could thus be greatly magnified.

WHERE DID THE INVENTION OCCUR?

When, where, and by whom the last remaining steps were taken in converting some ancestral form of the puttee into the *tipití* is difficult to say. The Witoto, Kofán, and Siona came close, but had not actually crossed the Rubicon. However, by 1909, the time of Whiffen's expedition to southern Colombia, the Bora, Andoke, Resigero, and Okaina, neighbors of the Witoto, were already using the *tipití* (Whiffen, 1915:132). Conceivably, the invention might have occurred in this region, the ancestors of one of these tribes having taken the last remaining steps. However, another interpretation seems more likely. It is that the *tipití* was invented well to the east of this region—probably somewhere in the Guianas, as Métraux (1928:104) surmised—and that it reached the Bora, Andoke, and other neighbors of the Witoto by diffusion. The spread of this device, moreover, might have been slow enough so that it had not quite reached the Witoto, Kofán, or Siona.[16] If they *had* been exposed to it (as was probably the case, given their proximity to their *tipití*-using neighbors), the advantage of the *tipití* over the puttee may not have been sufficiently obvious to have led them to abandon their traditional device.

Occurring near the western edge of Amazonia, the Witoto, Kofán, and Siona puttee may represent one of the few remaining survivals of a primitive device ancestral to the *tipití*, a device that once may have been much more widely distributed. The same may have been true of the more primitive Kayapó torsion sack, which is still found beyond the southern limits of the *tipití*.

We can summarize the matter as follows. Because the cultivation of bitter manioc is more widespread than the use of the *tipití*, the former is no doubt the older of the two. Because bitter manioc must be detoxified before it can be eaten, methods for doing so must have been hit on early in the game and become widely established. However, once a highly efficient method of detoxification—the *tipití*—was invented, it began to displace earlier and simpler methods, spreading out in all directions from its center of origin. Despite the advantages of the *tipití*, however, not enough time had elapsed since its invention for it to have replaced earlier squeezing devices everywhere. It should not be surprising, then, that vestiges of these earlier devices, including the *tipití*'s immediate precursor, should still be in use in outlying areas of Amazonia.

What we find here is a common pattern of diffusion. A device is invented that represents an improvement over previous devices used for the same purpose. From its point of origin, the invention spreads outward, and as it does so it supplants the more primitive forms when it comes in contact with them. Only beyond the margins of the distribution of this latest invention do simpler and less efficient devices manage to survive.[17]

In our attempt to reconstruct the evolution of the *tipití,* we are thus fortunate to encounter tribes like the Witoto, Kofán, and Siona, which, living just beyond the area to which the full-fledged *tipití* had spread, managed to preserve a more primitive form of manioc squeezer. This device—the puttee—is thus a "living fossil," close enough to the final form of the *tipití* to allow us to make a reasonable conjecture as to how the latter took its final evolutionary step. Had the Witoto, Kofán, and Siona already adopted the *tipití*, and had the puttee thus become extinct, establishing the chain of events leading to the *tipití* would have been much more difficult.

LAST STEPS IN THE INVENTION

The story does not quite end there. Another step remained to be taken before the full-fledged *tipití* could be said to have arrived. This step involved a change in the *material* used in making it. As we have seen, the Witoto and Kofán plait their puttees from strips of inner bark. The *tipití*, however, is plaited from narrower, stiffer, smoother strips usually obtained from the long stems of a reed (*Ischnosiphon* sp.), called *itiriti* in the Guianas and *arumã* in Brazil (Le Cointe, 1947:53–54; Roth, 1924:137–139).[18] These strips provide a tougher material, which lasts longer and is thus more appropriate for a device that, unlike the puttee, is meant to be used many times over. The transition from the puttee to the *tipití* thus was not complete until this change in materials took place. The use of this new material in ~~plaiting the *tipití* apparently did not occur immediately~~ or ~~automatically follow~~ing its invention. By a fortunate happenstance, however, considerable light can be shed on this transition.

THE TCHIKÃO *TIPITÍ* CLOSES THE RING

The schematic evolution of the *tipití* presented above had already been largely envisioned as early as 1960. However, it received a startling and unexpected confirmation in 1975 when, by chance, I saw in operation a *tipití* used by the Tchikão, a group of Indians newly arrived in the Upper Xingú. At the invitation of the French ethnologist Patrick Menget, who was studying the Tchikão, I was able to visit their village in the Xingú National Park. There, in one of their houses, I suddenly beheld a *tipití* like no other ever reported (Figure 11). Instead of being stretched *vertically*, as other *tipitís* invariably are, this one was stretched *diagonally.* The loop at the upper end of the Tchikão *tipití* was hooked over the end of a post inside the house, as is normally the case, but the lower end, which was plaited into a long tail like that of the Witoto puttee, was wrapped a turn and a half around a heavy log lying horizontally beneath it. The weight of this log, increased at times by having someone sit on top of it, stretched the *tipití.*

This manner of elongating the tube—diagonally, as with the Witoto and Kofán puttee—was indeed striking. Even more striking was the fact that the Tchikão *tipití* was plaited, not from reed splints at all, but from reddish-brown bast strips, indistinguishable to the eye from those used by the Witoto to make their puttee. Moreover, the Tchikão name for their *tipití—ponmu—*was the same as that for the tree from which they obtained the material for its manufacture. The inference seemed inescapable, namely, that this material—bast strips—was used in the very earliest *tipití*, before another material was discovered to be more suitable for its construction. In other words, what we find among the Tchikão is a *retention* of an older material, the kind of material traditionally used in the ancestral puttee, even after it had evolved into the first *tipití.*

The most likely interpretation of the fact that the Tchikão still possess this unique *tipití* seems to be as follows. Centuries ago, while still living somewhere near the Amazon River, the ancestors of the Tchikão were introduced to and adopted the newly invented *tipití*. Being then in its earliest stages, the *tipití* was made from the same material formerly used for its precursor. Just about this time—that is, before reed splints had replaced bast strips in making the *tipití*—the Tchikão began their southward trek. Having moved beyond the range of the *tipití*, they were never exposed to the new improvement made in its construction and never hit on it themselves.

There is yet one more thing to observe about the Tchikão *tipití*, namely, the manner in which it was plaited. Unlike the Witoto and Kofán puttees, which were plain plaited (over-one, under-one), the Tchikão *tipití* was twill plaited (over two, under-two), an apparent advance in plaiting technique shared with other *tipitís*. Twill plaiting the body of the *tipití* (which occurs in a variety of different patterns, as shown in Roth [1924:facing 283; see also Falco et al., 1987:151]), was probably hit on as a way of imparting more flexibility to the *tipití*, thus giving it a greater

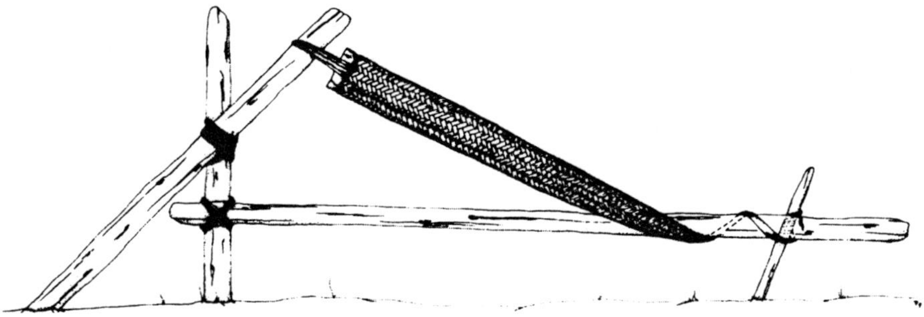

Figure 11. Tchikão *tipití* showing diagonal stretching of the device. It is twill plaited on the bias for greater flexibility in squeezing the contents.

ability to compress the enclosed pulp.[19] The fully evolved *tipití* is, in fact, a rather intricately plaited device, whose manufacture has been described in detail by Roth (1909:27–31, 1924:282–286, 1929:30–31). However, a consideration of the details exhibited by later forms of the *tipití* lies outside the scope of this chapter.

In summary, we can say that the Tchikão, like the Witoto, Kofán, and Siona, have contributed in no small measure to our understanding of how the *tipití* gradually evolved.

RECONSTRUCTED STAGES OF DEVELOPMENT

With substantial ethnographic evidence on which to draw, accompanied by plausible conjectures, we have been able to piece together the probable evolution of the *tipití*, from its simple beginnings to its final form. The stages of this development may be summarized as follows: (1) squeezing grated manioc pulp with the hands to express its poisonous juice; (2) squeezing it with a large leaf; (3) squeezing it inside a piece of bark cloth; (4) squeezing it by means of a spirally wrapped bark strip; (5) squeezing the grated pulp inside two sets of attached palm-leaf midribs that are then twisted; (6) squeezing the pulp by means of torsion inside a flexible plaited sack (the Kayapó torsion sack); (7) squeezing it by means of a plaited web of bast strips (the puttee) wrapped spirally around manioc pulp and held horizontally, sticks being inserted through loops at each end and twisted in opposite directions; (8) plaiting pockets at each end of the puttee, thus taking the first step toward weaving it as an enclosed tube; (9) hanging the puttee from the end of a beam and squeezing the contents of the tube thus formed by pulling it out diagonally and twisting it with a stick through a loop at the lower end; (10) squeezing the manioc pulp in a cylindrically plaited tube (the *tipití*), still made of the same bast strips as used in the puttee, and stretching it diagonally; and (11) the *tipití* in its final form, being plaited with splints cut from a reed and hung vertically, the tube being stretched by means of a heavy weight or a lever.

CONCLUSION

I have proposed a succession of forms that led from initially simple means for detoxifying bitter manioc to the sophisticated and efficient device known as the *tipití*. The gaps between each stage—and thus the leap required to bridge them—do not appear insurmountable. On the contrary, each form in the proposed succession of stages seems to come readily out of the preceding one, incorporating certain elements of it.

That the sequence of events depicted here should consist of a number of easy gradations exemplifies the history of virtually all great inventions, namely, that they occur, not through one remarkable saltation, but by the gradual accumulation of relatively small incremental advances. Once the steps in the process have been identified and described, the end result to which they lead—the ultimate synthesis—is shorn of much of its mystery.

Amazonian Indians once faced a serious problem: how to detoxify a poisonous tuber that could then be made into a valuable food. They first went about solving this problem by very simple means, but then gradually elaborated and improved on these means. From this vantage point, the need to call on extraordinary "genius" to account for the origin of the *tipití* disappears. The invention proves to be, not a single extraordinary event, but a protracted and discernible *evolution*. At no point did it require more than the deliberate application of normal human intelligence, working with the knowledge that there was a problem to be solved and having familiarity with the materials and techniques needed to arrive at successively better solutions.

The case history of the *tipití* is thus no exception to the rule that apples to scientific inventions generally. Indeed, it stands as one more confirmation of this rule. The invention of the *tipití* was, to be sure, a significant cultural advance, but this advance, and others like it, is better understood as the working out of an intelligible process of cultural development than as an unaccountable flash in the mind of some unknown primitive genius.

ACKNOWLEDGMENTS

This chaper was first conceived as a joint project by the late Anthony Leeds, Gertrude E. Dole, and me. After a promising start, it was set aside and remained untouched for many years. Only now, with an invitation to contribute to this volume, has it received the impetus necessary to bring it to a conclusion. During the last phase of the work, the suggestions and research assistance of Laila Williamson were most helpful. Finally, although his contributions to the project are acknowledged in the appropriate place, I would like to make particular note here of my debt to the work of Borys Malkin, whose field observations, ethnographic collections, and photodocumentation contributed immeasurably to the reconstruction presented in this chaper.

NOTES

1. *Tipití* is the Tupí-Guaraní name commonly employed in Brazil for this device. In the literature, the use of this term goes back at least to 1557, when it was mentioned in the first edition of Hans Staden's *The True History of His Captivity* (1928:130). In the Guianas, the device is generally known as a *matapi* and in Venezuela and the West Indies as a *sebucaán* or *cibucán*.
2. Among some tribes, a second vertical post, with notches cut up one side, is placed into the ground opposite the first post. As the squeezing proceeds, the free end of the lever is inserted into successively lower notches of the second post, stretching the *tipití* even farther and thus exerting greater pressure on the contents.
3. When extended longitudinally, the interior volume of one small *tipití* from the Vaupés region I measured decreased from 1,770 cu cm (108 cu in) to 826 cu cm (50.4 cu in). Audrey Butt Colson found the maximum diameter of a large Akawaio *tipití* to be 16.5 cm (6.5 in) when compressed from the ends, its minimum diameter decreasing to12 cm (4.7 inches) when fully extended (Colson, personal communication, 1960).
4. In a related issue, it is still a matter of speculation whether manioc tubers were originally poisonous, or whether generations of selection for higher starch content have had the collateral but unintended effect of increasing the prussic acid content as well. Either way, at whatever stage in the domestication of manioc its prussic acid became a serious problem, the impediment was confronted and overcome.
5. The tribes of this region grow only bitter manioc and in ridding it of its poison use a mat strainer made of palm-leaf midribs twined together with wild pineapple fiber or cotton thread. The device is first laid on top of three slats resting on a large flat-bottomed pot. The grated manioc pulp is laid on top of the strainer and pressed against it with the hands, the poisonous juice thus squeezed out falling to the bottom of the pot below. However, because this device does not seem to lie on the direct evolutionary line leading to the *tipití*, it will receive no further consideration here.
6. This specimen is in the collections of the American Museum of Natural History and bears the catalog number 40.1/125. The *hanto* is a sieve plaited on the bias in concave form from palm-leaf midribs. Unlike the *tipití*, it is used to sift dry manioc flour rather than to strain wet pulp.
7. It was also found that by slicing the manioc root and drying it in the sun, about 75% of its prussic acid content was dispelled (Chadha, 1961:106). The only instance I am familiar with of the use of this method by South American Indians is provided by the Mojo of eastern Bolivia, among whom Métraux (1948:413) reported: "Bitter manioc tubers were sliced thin and dried in the sun. . . ." However, Koster (1816:370) observed the use of this method by Caboclos in northeastern Brazil more than a century and a half ago, and it is possible that it may have been borrowed from neighboring Indians.
8. For a detailed account of the intricacies involved in detoxifying poisonous manioc tubers, both before and after grating and before and after pressing, see Dufour (1995:156).
9. In the absence of a thorny palm root, manioc can be grated on a rock, as the Waica (Yanomamö) do (Barker, 1953:450) and the Tupinambá are reported to have done (Staden, 1928:137). If a mortar and pestle are available, they can also be used to mash up manioc roots, as the Amahuaca sometimes do.

10. This device was no longer in general use among the Sherente at the time they were studied by David Maybury-Lewis, having been replaced by the *tipití*. However, the Sherente were still familiar with it and made one (from bast strips instead of palm leaves) and demonstrated its use for the ethnographer (Maybury-Lewis, personal communication, 1962). Galvão (1963:131) called the device a *tipití de torção*, but it is misleading to use "*tipití*" as a generic term for manioc squeezer instead of as the name of particular type of squeezer.
11. Moreover, William T. Vickers, a recent student of the Siona, reports their manioc squeezer as being plaited (rather than consisting of a solid sheet of bark cloth) and as operating by means of a short stick tied to the lower tail of the device and then twisted, tourniquet fashion, just as Hardenburg described three-quarters of a century ago (Vickers, personal communication, 1996).
12. In an experiment using a Witoto puttee from the collections of the Museu Paulista in São Paulo, Aytai (1984:120) found that, when twisted to the maximum with manioc pulp inside, the interior volume of the device was compressed to about 32% of its initial capacity.
13. Photographs in both Whiffen and Tessmann reveal that early in this century the Witoto puttee was already plaited on the bias. This technique is surely an old one in Amazonia, for it results automatically when a basket is being plaited from a single palm leaf, as is done by many tribes, such as the Sherente (American Museum of Natural History [AMNH] specimen 40.0/9901). Thus it must have been part of an existing substratum of basket-making techniques, ready to be applied when experimentation with manioc pressing began.
14. The Witoto also employ an elaboration of this method. Close to the beam from which the upper end of the puttee is hung, a vertical post is set into the ground and used as a fulcrum around which to twist the stick that has been inserted through the lower loop of the puttee. The stick may be turned as much as 180°, producing considerable pressure on the contents (Figure 8).
15. This juice contains in suspension the fine starch flour, tapioca, and thus, if it is thrown away without any further treatment, a useful nutritional element is lost.
16. Neither had it reached the tribes of the Orinoco and the Upper Río Negro. Thus Roth (1924:281) wrote: "From the fact that Gumilla, so accurate an observer, makes no mention of a matapi [*tipití*] in his description of cassava manufacture, though he speaks of the poisonous juice being expressed . . . , it may perhaps be assumed that the article was not known in his day on those portions of the Orinoco with which he treats. This conclusion is quite feasible, considering that even at the present day on the Uaupes River, not so very far away from it, the juice is still expressed by hand pressure" (Roth, 1924:217). More precisely, it is pressed with the hands through a circular strainer supported on a triangular frame, a receptacle being placed below it to catch the juice.
17. This is also an exemplification of the age-area concept, according to which the older forms of a trait are found at the margins of its distribution, having been displaced near the center by newer and presumably superior forms.
18. However, Richard Schomburgk (1922:95) noted that the Warao *tipití* was made from another reed, a species of *Calathea*, although Farabee (1924:21) reported that the Makushí *tipití* was plaited from splints cut from the fronds of the royal palm, *Maximilliana regia*. *Ischnosiphon* splints may be 2.5 or 3 m (9 or 10 ft) long (Roth, 1909:27)—long enough to spiral nearly the full length of a *tipití*.
19. In one *tipití* I examined, when the tube was compressed longitudinally, as is normally done in filling it with manioc pulp, the opposed sets of plaiting elements crossed each other at an angle of 125°, substantially more than a right angle. However, when stretched to the limit, the opposed sets of plaiting elements, "scissoring" past each other, reduced this angle to 55°, indicating how much "play" twill plaiting permits the *tipití* to have. This degree of play in a working *tipití*, as Roth (1929:31) noted, "allows of more elasticity on being filled and emptied."

REFERENCES

Aytai, D., 1987, O *Tipití*-de-torção. *Publicaçáes do Museu Histórico de Paulínia* 35:110–122.

Barker, J., 1953, Memoria sobre la cultura de los Guaika. *Boletín Indigenista Venezolano* 1:433–489.

Bennett, W. C., and Bird, J. B., 1949, *Andean Culture History*. Handbook Series No. 15. American Museum of Natural History, New York.

Carneiro, R. L., 1953, Field Notes. Department of Anthropology, American Museum of Natural History, New York.

Chadha, Y. R., 1961, Sources of Starch in Commonwealth Territories. III: Cassava. *Tropical Science* 3:101–113.

Cook, O. F., 1903, Food Plants of Ancient America. *Smithsonian Institution Annual Report* 1903:481–497.

Cooke, R., and Cock, J., 1988, Cassava Crops Up Again. *New Scientist* June 17:63–68.

Dole, G. E., 1960, Techniques of Preparing Manioc Flour as a Key to Culture History in Tropical America. In *Selected Papers of the Fifth International Congress of Anthropological and Ethnological Sciences*, pp. 241–248. University of Pennsylvania Press, Philadelphia.

Driver, H. E., 1961, *Indians of North America*. University of Chicago Press, Chicago.

Dufour, D. L., 1995, A Closer Look at the Nutritional Implications of Bitter Cassava Use. In *Indigenous Peoples and the Future of Amazonia*, edited by L. E. Sponsel, pp. 149–165. University of Arizona Press, Tucson.

Ehrenreich, P., 1948, Contribuçáes para a etnologia do Brasil. *Revista do Museum Paulista* 2,n.s.:7–135. (Originally published in 1891.)

Falco, J. R., Pazinatto, R., and Aytai, D., 1987, Tipití—contribução a seu estudo. *Revista do Museu Paulista* 32(n.s.):131–153.

Farabee, W. C., 1924, *The Central Caribs*. Anthropological Publications Vol. 10. University Museum, University of Pennsylvania, Philadelphia.

Frikel, P., 1968, *Os Xikrín*. Publicações Avulsas No. 7. Museu Paraense Emilio Goeldi, Belém, Brazil.

Galvão, E., 1963, Elementos básicos de horticultura de subsistencia indígena. *Revista do Museu Paulista* 14,n.s.:120–144.

Hardenburg, W. E., 1912, *The Putumayo, The Devil's Paradise*. T. Fisher Unwin, London.

Heizer, R. F., and Elsasser, A. B., 1980, *The Natural World of the California Indians*. University of California Press, Berkeley and Los Angeles.

James, W., 1917, Great Men and Their Environment. In *Selected Papers on Philosophy*, pp. 165–197. Everyman's Library Edition. J. M. Dent and Sons, London.

Koch-Grünberg, T., 1910, *Zwei Jahre unter den Indianern: Reisen in Nordwest-Brasilien, 1903–1905*. Volume 2. Ernst Wasmuth, Berlin.

Koster, H., 1816, *Travels in Brazil*. Longman, Hurst, Rees, Orme, and Brown, London.

Krause, F., 1911, *In den Wildnissen Brasiliens*. R. Voightlander Verlag, Leipzig.

Le Cointe, P., 1922, La culture et la préparation du manioc en Amazonie. *Revue de Botanique Appliquée et d'Agriculture Coloniale* 2:331–337.

Le Cointe, P., 1947, *Arvores e plantas uteis*. Biblioteca Pedagógica Brasileira, Série 5, Brasiliana, Volume 251. Companhia Editora Nacional, São Paulo.

Lévi-Strauss, C., 1948, The Nambicuara. In *Handbook of South American Indians, Vol. 3: The Tropical Forest Tribes*, edited by J.H. Steward, pp.361–369. Bureau of American Ethnology Bulletin No. 143. Smithsonian Institution, Washington, D.C.

Lipkind, W., 1949, The Carajá. In *Handbook of South American Indians, Volume 3: The Tropical Forest Tribes*, edited by J. H. Steward, pp. 179–191. Bureau of American Ethnology Bulletin No. 143. Smithsonian Institution, Washington, D.C.

Lowie, R.H., 1946, The Northwestern and Central Ge. In *Handbook of South American Indians, Volume 1: The Marginal Tribes*, edited by J. H. Steward, pp. 477–517. Bureau of American Ethnology Bulletin No. 143. Smithsonian Institution, Washington, D.C.

Lowie, R. H., 1948, The Tropical Forest: An Introduction. In *Handbook of South American Indians, Volume 3: The Tropical Forest Tribes*, edited by J. H. Steward, pp. 1–56. Bureau of American Ethnology Bulletin No. 143. Smithsonian Institution, Washington, D.C.

Malkin, B., 1975a, Cofán (South Colombia, Montaña) Plaiting a Manioc Press "Tipiti." In *Encyclopaedia Cinematographica*, edited by G. Wolf, pp. 1–12. Institut für den Wissenschaftlichen Film, Göttingen.

Malkin, B., 1975b, Cofán (South Colombia, Montaña) Manioc Processing. In *Encyclopaedia Cinematographica*, edited by G. Wolf, pp. 1–12. Institut für den Wissenschaftlichen Film, Göttingen.

Malkin, B., 1975c, Witoto (East Peru, Selva) Plaiting a Manioc Press "Tipiti." In *Encyclopaedia Cinematographica*, edited by G. Wolf, pp. 1–10. Institut für den Wissenschaftlichen Film, Göttingen.

Malkin, B., 1975d, Witoto (East Peru, Selva) Manioc Processing. In *Encyclopaedia Cinematographica*, edited by G. Wolf, pp. 1–10. Institut für den Wissenschaftlichen Film, Göttingen.

Métraux, A., 1928, *La civilisation matérielle des tribus Tupí-Guaraní*. Librairie Orientaliste Paul Geuthner, Paris.

Métraux, A.,1948, Tribes of Eastern Bolivia and the Madeira Headwaters. In *Handbook of South American Indians, Volume 3: The Tropical Forest Tribes*, edited by J. H. Steward, pp.381–454. Bureau of American Ethnology Bulletin No. 143. Smithsonian Institution, Washington, D.C.

Milham, W. I., 1947, *Time and Timekeepers*. Macmillan, New York.

Nimuendajú, C., 1939, *The Apinayé*. Anthropological Series No. 8. Catholic University of America, Washington, D.C.

Nordenskiöld, E., 1929, *Modifications in Indian Culture through Inventions and Loans*. Comparative Ethnographical Studies No. 8. Göteborg.

Oberg, K., 1953, *Indian Tribes of Northern Mato Grosso, Brazil*. Institute of Social Anthropology, Publication No. 15. Smithsonian Institution, Washington, D.C.

Oviedo y Valdés, G. F. de, 1851, *Historia general y natural de las Indias*, Volume 1, Book 7, Chap. 9, edited by J. Amador de Ríos. Imprenta de la Real Academia de la Historia, Madrid.

Ribeiro, B. G., 1988, *Dicionário do artesanato indígena*. Editora Itatiaia, Editora de Universidade de São Paulo, Belo Horizonte.

Rogers, D. J., 1965, Some Botanical and Ethnological Considerations of *Manihot esculenta*. *Economic Botany* 19:369–377.

Roth, W. E., 1909, Some Technological Notes from the Pomeroon District, British Guiana, Part I. *Journal of the Royal Anthropological Institute* 39:26–34.

Roth, W. E., 1924, *An Introductory Study of the Arts, Crafts, and Customs of the Guiana Indians*. Thirty-eighth Annual Report of the Bureau of American Ethnology, 1916–1917. Government Printing Office, Washington, D.C.

Roth, W. E., 1929, *Additional Studies of the Arts, Crafts, and Customs of the Guiana Indians*. Bureau of American Ethnology Bulletin No. 91, Smithsonian Institution. Government Printing Office, Washington, D.C.

Sauer, C. O., 1950, Cultivated Plants of South and Central America. In *Handbook of South American Indians, Volume 6: Physical Anthropology, Linguistics and Cultural Geography of South American Indians*, edited by J. H. Steward, pp. 487–543. Bureau of American Ethnology Bulletin No. 143. Smithsonian Institution, Washington, D.C.

Schomburgk, R., 1922, *Richard Schomburgk's Travels in British Guiana, 1840–1844*, Volume 1. Translated and edited by W. E. Roth. Daily Chronicle Office, Georgetown, British Guiana.

Sobel, D., 1995, *Longitude*. Penguin Books, New York.

Spinden, H. J., 1917, The Origin and Distribution of Agriculture in America. In *Proceedings of the 19th International Congress of Americanists*, pp. 269–276. Washington, D.C.

Staden, H., 1928, *The True History of His Captivity, 1557*. George Routledge and Sons, London.

Tessmann, G., 1930, *Die Indianer Nordost-Perus*. Friederichsen, De Gruyter, and Company, Hamburg.

Tylor, E. B., 1871, *Primitive Culture*. 2 vols. John Murray, London.

Vázquez de Espinosa, A., 1942, *Compendium and Description of the West Indies*. Translated by C. U. Clark. Smithsonian Miscellaneous Collections, Volume 102, Washington, D.C.

Wagley, C., 1977, *Welcome of Tears*. Oxford University Press, New York.

Watt, G., 1908, *The Commercial Products of India*. John Murray, London.

Whiffen, T., 1915, *The North-West Amazons*. Constable and Company, London.

White, L.A., 1948, *The Science of Culture*. Farrar, Straus, and Company, New York.

Worsley, P., 1962, The Utilization of Natural Food Resources by an Australian Aboriginal Tribe. *Acta Ethnographica Academiae Scientiarum Hungaricae* 10:153–190.

Chapter **4**

Accelerated Technological Change in Archaeology and Ancient History

ROBERT McC. ADAMS

INTRODUCTION

Discontinuous, sometimes violent change is universal. In the short term, it threatens all living systems, even if in the long term transformative change may offer the best or only hope of assuring their successful adaptation to new conditions and hence survival. This applies as much to human as to natural history. In this chapter, I discuss three instances of one type of discontinuous change that serves as a conceptual tool for archaeological and historical research on ancient societies.

Built into traditional Darwinism, "descent with modification" was an acceptance of the standpoint of Lyell's geological gradualism. In its time, his assumption of uniformitarian change in the earth's geological history carried the day against competing doctrines of catastrophism. Today, however, there is increasing recognition of great diversity in rates of evolutionary change. This includes, but is not limited to, repeated episodes of abrupt and catastrophic climatic change that contributed to periodic widespread extinctions and consequent major restructurings of the biosphere. Subsequent evolutionary directions, in no way immanent in the

ROBERT McC. ADAMS • Department of Anthropology, University of California, San Diego, La Jolla, California 92093.

Cultural Evolution: Contemporary Viewpoints, edited by Gary M. Feinman and Linda Manzanilla. Kluwer Academic/Plenum Publishers, 2000.

extinctions themselves, can in retrospect be seen to have created the conditions for newly emergent levels of greater complexity. Formerly marginal groups, moving into newly vacated niches, proliferated and diversified. It has recently been suggested that in those new niches they also may have been enabled to evolve more rapidly (Damgaard, 1996; Fitch and Ayala, 1995).

Accelerated phases of change of this kind, often referred to in evolutionary biology as punctuations (Eldredge and Gould, 1972; Gould and Eldredge, 1993; cf. Mayr, 1988), invite closer study by students of human as well as natural history. Granting that the mechanisms involved in generating and selectively adopting changes in biological and cultural evolution are different, it is widely agreed that culture, too, can usefully be regarded as behaving over time like a kind of living system (Spencer, 1990:4). A common, fundamental feature in both kinds of systems is the critical importance of relatively brief, transformative episodes. As Marcus and Flannery remarked,

> periods of transition, or rapid evolution, are more important than periods of stability. Evolutionary theory has often been criticized for focusing too heavily on stable periods of slow evolution, which form the basis for stages like autonomous village, chiefdom, and so on (1996:244).

Whether there ever were periods of prolonged, complete cultural stability seems doubtful. Both stability and the notion of evolutionary "stages" that is largely based on the assumption of its existence need to be recognized as essentially artificial constructs. At least during the later evolution of the more complex and interconnected forms of human society, intensified external contacts and internal variability seem likely to have sustained some level of virtually continuous change. Even in Pleistocene times, the risks and stresses associated with ever-present ecosystemic variability are also likely to have selected for more or less continuous adaptive modification if not directional, evolutionary change in some cultural features (Potts, 1996). The assumption of stability bears some similarity to the no less doubtful—if for a time useful as a platform for formal theory construction—assumption of equilibrium on which much of classical economics was originally erected.

But the reality of great irregularity in rates of change, interspersing slowly evolving conditions with greatly accelerated transitional phases, is not similarly in doubt. As has been repeatedly and convincingly demonstrated for recent centuries of technological change in particular, successive, mutually reinforcing clusterings of major innovative events have come together to constitute larger, transformative impulses (see most recently Girifalco, 1991). Building on the original insights of Schumpeter (1939), these waves or clusterings form the basis of relatively rapid changes in the fundamental direction of general industrial growth.

Differences remain, and are not easily resolved, in establishing the precise timing and most significant features of these clusterings. There is always a subjective element in deciding on relative degrees of importance of innovations and in establishing the moment of their origin in what is usually a complex and fairly protracted series of discoveries. But both the irregularity of rate and the "path-dependent" (i.e., dependent on historical context, not teleologically determinate) character of the process are beyond controversy.

We frequently characterize the most salient of the human subset of accelerated changes as revolutions. Clearly, this subset has features not found in its natural counterparts. As a species, *Homo sapiens* is far more capable than most of generating internally its own change-inducing stresses. We need no external stimuli in the form of natural catastrophes to start our revolutions. Largely accounting for this are what can be described as our species' emergent capacities for individual as well as group goal orientation and planning, for cumulatively retained learning, and for the enormously enhanced communication that only human language permits.

Punctuated episodes of change have not occurred only at long, effectively random, or externally imposed intervals in human history. Very widely spaced early in the human career, the intervals between them began to shorten appreciably in the Late Pleistocene. Since the beginnings of agriculture, they have recurred at a gradually accelerating rate. With modernity, revolutionary clusterings of change in different sectors may quickly succeed one another or even overlap in time.

Major episodes of discontinuous, accelerated change form, at best, a somewhat open-ended category. Political revolutions, relatively brief and frequently accompanied by violence, lie at one end of a continuum. With considerable popular support for organized courses of abrupt action as a requisite, their failure or success (seldom the same as the one originally envisioned) is generally decided within a few years at most. The Industrial Revolution is perhaps the prototype of more moderately paced changes at the other end of the continuum. An accelerated rate of change and a cumulatively transformed outcome are once again a defining requirement. But one or even several generations may pass before the full significance of what has occurred is widely recognized by participants. A goal-oriented, consciously planned program of comprehensive action by a committed segment of the society may be altogether missing.

The abruptness and short duration of politically motivated revolutions make it difficult to enrich textually based accounts of them with archaeological findings. Archaeological chronologies are inherently imprecise. Those relying exclusively on radiocarbon dating are reducible to plus-or-minus intervals of appreciably less than a century only in special circumstances and with considerable effort. Hence archaeological interest has been primarily directed toward more gradualistic forms of change. There is a tendency to ignore, or to dismiss as beneath the threshold of "scientific" interest, changes for which chronological matrices based on the archaeological record alone are too coarse grained to capture.

There are exceptions, to be sure. The relatively accurate dendrochronological dating of burned building remains, or the attestation of a relatively contemporary phase of widespread abandonment through settlement pattern surveys, may coincide with textual references to a disaster. Conversely, a royal reign whose annals boast that it was marked by monumental building programs and expansionism can usually find archaeological confirmation as well. If a transitional period of accelerated change lasts long enough, microseriation or careful stratigraphic control may permit archaeologists to determine whether it had a fairly abrupt, stepwise, or a more gradualistic, ramplike form.

However, as these limited possibilities indicate, it is rare for archaeological investigations to add much to an understanding of processes of change of a brief and primarily political character. Event-centered changes may escape archaeological recognition altogether even where they have previously been unambiguously attested in textual records. Consequently, the myopia of ancient written sources, dominated by the viewpoints and activities of relatively narrow elites (and sometimes only of factions in them), imposes obvious biases. It is usually difficult or impossible with ancient documentary sources alone to ascertain the wider interaction of social and economic forces responsible for change and innovation in society as a whole.

The irony is that this is precisely the contribution that—at least in the abstract—archaeololgy is best qualified to make. Typically, however, "civilizational" archaeologists have until fairly recently paid relatively little attention to ordinary living quarters, workshops, and outlying, small settlements. With interests centering on monumental buildings and the recovery of texts and works of artistry, their preoccupation long paralleled the myopia of the ancient scribes with "dynasties, wars, and religions." This further intensifies an elitist bias.

Still another source of bias is to be found in early Near Eastern textually oriented studies. Most archival sources provide what may be figuratively described as narrow shafts of light in an immense, dark room. Texts generally refer to only a restricted range of actors and behaviors; long, continuous chronological coverage is exceptionally rare. The natural tendency is to connect these small points of illumination with continuous lines. This helps enormously if the aim is merely to draw a reassuringly coherent picture. But no less than in archaeology, historiographic reconstructions are thus led implausibly to deny the virtual certainty of many deep discontinuities.

What can be done to correct the picture? One important opportunity focuses on technological as opposed to political change. This, too, can take the form of relatively rapid, sustained transitions of a cumulatively "revolutionary" character, as is illustrated by prevailing references to the Industrial Revolution of the late 18th and early to mid-19th centuries in Britain. Without suggesting the direct comparability of these occurrences, a broadly similar focus on technology in the ancient world can compensate for the elitist bias inherent in many of our traditional sources and approaches. It thus may provide a more balanced view of the full array of forces for change that were at work in ancient society.

TECHNOLOGICAL ACCELERATION IN THE FIRST URBAN SOCIETIES

The Mesopotamian manifestation of the beginnings of civilization, consummated toward the end of the fourth millennium B.C., provides the earliest known example of state-level societies. It was long ago characterized by Childe (1950) as an urban revolution, this interpretation stressing both the magnitude and the accelerated rate of the associated changes. Over a span of at most several centuries for the key innovations involved—following, it must be remembered, 5,000 to 7,000 years of much more slowly developing village and town life—a number of key technological and organizational features were more or less concurrently introduced or significantly increased in scale and complexity.

Bringing together these changes under the general characterization of a specifically *urban* revolution, as Childe did, is now admittedly debatable. On present evidence, new or greatly enlarged settlements of urban size did indeed provide a kind of spatial envelope in the Mesopotamian case in which most or all the known changes were concentrated. However, this does not necessarily indicate that the crossing of some threshold of settlement size (even if there were dependent, closely related organizational changes) was the decisive, explanatory factor in the emergence of civilization. We are, however, justified in concluding that cities were, in most instances, at least the primary locus of change, as certain key industries were the central loci of change in the Industrial Revolution. The more widely accepted alternative to applying a general characterization of urbanism to the transformation as a whole identifies the emergence of new state superstructures as the decisive factor. But although much evidence can be adduced in support of this latter approach, I believe its limitation of focus to the same elitist sources of bias mentioned earlier has not received the attention it deserves.

For the subject matter of technology to serve our wider purpose, more is clearly entailed in its definition than merely tools and techniques. More accurate and useful is a definition extending it to all "knowledge of how to fulfill certain human purposes in a specifiable and reproducible way . . . something that can be reconstructed in principle through specifiable algorithms." No less important, as Brooks went on to observe, it also "must be sociotechnical rather than technical, and [thus] . . . must include the managerial and social supporting systems necessary to apply it on a significant scale" (1980:65–66). Technology, in short, must be understood as deeply suffused with socioeconomic—and indeed cultural—meaning.

The core changes constituting the urban revolution and/or emergence of state societies in "greater Mesopotamia" have been reasonably familiar ever since Childe's early work. A recent symposium is a welcome overview not only of present controversies, but of the veritable transformation of the subject as a result of new data acquired over the last decade or two (Rothman, in press). It is now known, in particular, that an *oikumene* of cultural and economic interaction spanned an enormous area already for much of the fourth millennium B.C. This extended at least from the Mediterranean coast to the central Iranian Plateau, and from the highlands of Anatolia to the head of the Persian Gulf.

Binding this vast region together, apart from reciprocal trade relations, were population movements of substantial but still undetermined extent and character that seem to have emanated from southern Mesopotamia. In the course of these movements, urbanism, manifested earliest and most extensively there, spread dramatically. Still unresolved and subject to vigorous debate, as the symposium testified, are the character and extent of the accompanying political integration that these movements imply.

For present purposes, I call attention to only a few of the most salient features of structural change, stressing their interrelated, systemic character. Compared with earlier segmentary societies rather loosely designated as "chiefdoms," the new urban organizations were significantly more stratified. This is architecturally symbolized by monumental public building; "temples" and "palaces" are at first only somewhat conventionally distinguishable, although they must have quickly begun to take on differentiated functions as well as formal features. Each city and its temple complexes was viewed as the residence and supporting establisment of one or more sponsoring deities.

Glyptic representations of military activities and (more questionably, contemporary) myths of hostilities projected into the world of the gods suggest the increasing prominence of warfare and show that war leadership played a large part in the emergence of new forms of sociopolitical leadership. We may never know whether a formerly more dispersed population came together on a primarily voluntary basis to build cities or was compelled to do so by newly emergent forms of increasingly militaristic, repressive leadership favoring this concentration for its own purposes. City construction is in any case likely to have entailed heavier and more regimented forms of resource extraction and corvée labor for defensive and other public works than had previously been known. Walled settlements that were for the first time urban in character and scale offered an additional margin of protection for food reserves as well as for their inhabitants. Bases made secure in this manner also conveyed offensive advantages against smaller neighboring communities—a significant gain in a semiarid region beset by great variability in the productiveness of its soil, water, and other resources.

There is no reason to doubt that politico-religious leaders were also involved in setting technology's major directions along with many other social goals. An indisputable means of doing so lay in their ability to make provision for the support of specialists whose products satisfied elite institutional obligations and personal appetites. But the primary responsibilities of these elites were, after all, to direct various forms of concerted action, to administer the redistribution of resources, and to represent the community before the gods in public ceremonies. It is most unlikely that they also should be assigned major credit for the extraordinary outpouring of creativity that ensued in mid- to late fourth millennium B.C. Mesopotamia.

At least in its long-term implications, cuneiform writing on clay tablets was the most decisive of the many technical innovations that made their appearance at this time. Originally in pictographic form, it seems to have appeared rather suddenly and dramatically. Primarily serving from the outset the needs of economic administration in newly formed city-states, it quickly became vital in administering receipts and disbursements at state or temple storehouses. Over time, the originally pictographic forms were simplified and standardized into a mixed ideographic and syllabic system. Some centuries later, they began to approximate the grammatical structures of ordinary speech and thus to serve the wider purposes of substantive communication.

Before the advent of writing, there was a lengthy sequence of preceding forms of symbolic notation that clearly met gradually emerging needs for counting, recording, and communicating various kinds of receipts and disbursements. Writing itself was thus "an additional, albeit crucial step in a sequence of experiments in implementing effective control mechanisms in the economy." As Hans Nissen continued, "It can be adduced with some degree of confidence that the organizational structures of society and economy were already well established when literacy came into being" (Nissen et al., 1993:19, 21).

Signs testifying to the presence of status hierarchies and politico-religious leadership are found already in some of the earliest, still pictographic texts. So is evidence of an extensive, hierarchically specified division of labor, including diverse groupings of craft specialists working in wood, metals, and more exotic raw materials brought from great distance. In short, writing, crafts, and new forms of administrative organization emerged together, and rather rapidly. They appear to constitute a single complex, some combination of both stimulus and response to major changes in social scale and complexity.

It cannot be accidental that the appearance of writing coincides with the first appearance of large population aggregates. Food procurement necessarily became a more carefully scheduled, specialized, centrally controlled activity. Resources of labor were now in hand to extend and intensify local irrigation grids. More complex provisions were necessary for circulating secondary products of domestication and smoothing sectoral fluctuations in subsistence outputs. Calling for new forms of record keeping, in particular, was a newly instituted, centrally redistributed ration system. Monumental buildings signified the attainment of forms of organizational complexity as unprecedented as they were. The co-evolution of writing and urbanism represented a powerful, embracing—in my terms essentially technological—process that lent support to a more aggressively centralized and directed form of social integration.

Unfortunately, the full extent of the linkage between the many elements of urbanism is still obscure. We learn of little more than the activities of the "great organizations"—palaces and temples—from the textual sources and much of the available archaeological evidence as well. Areas of most immediate concern to these organizations and the elites controlling them, such as the long-distance procurement of costly raw materials, are most likely to have been carried on as centralized enterprises and hence to be at least relatively well attested. Craft specialists engaged in this work are listed in varying degrees of attachment to or dependence on the great organizations. But the myopia of our sources is such that we cannot exclude the parallel involvement of many of these same specialists, as well as many others and a corresponding flow of noncentrally administered resources, in a parallel, private economy whose written records (if any) rarely found their way into central archives (Stein and Blackman, 1993:50–55; Zettler, 1996:19–20).

Were craftsmen engaged by the great organizations, even though often on only a part-time basis, somehow kept apart from others largely engaged in more mundane production? That seems unlikely. Were the individually specialized crafts then organized along guild lines with common standards, modes of distribution, apprenticeship, and so on? The possibility of even modest tendencies in this direction invokes the further possibility of at least a degree of craft autonomy and casts doubt on the possibility of complete subservience to the new politico-religious elites. But how and to what extent did their mundane products, if not their artistry, meet the growing demands of the population at large? For the mass production of pottery, now at times employing the use of the potter's wheel, it seems clear that there was a virtual continuum of styles and standards from inside the great organizations to the larger communities around them. For the greater number of other crafts as well, although we cannot yet ascertain their patterns of distribution with any certainty (Postgate, 1992:229), it probably would be wise to resist the notion that palace or temple elites had complete control over either the output or the rapidly advancing skills of the crafts. As Stein observed,

> The distinction between attached and independent specialists should not be viewed as an absolute dichotomy, but rather as a characterization of two different ends of a continuum of variation in the degree of elite control over craft production. (1996:25–26)

Just as the introduction and evolution of writing testifies to a great creative surge, so do many if not all the other crafts. Although the existing archaeological record places undue emphasis on cultic objects of the late fourth millennium B.C., it is noteworthy that at least these iconic categories exhibit common qualities of technical precocity and high artistic creativity. As Frankfort once noted of the stone cylinder seals of the time, for example, they were endowed from their first appearance with a "creative power . . . such that we meet among its astonishingly varied products anticipations of every school of glyptic art which subsequently flourished in Mesopotamia" (1939:23).

The feature of these seals that is particularly significant for my purposes is that they played an integral part in the newly emergent technology of administration of their time. The basic techniques of carving and inscribing stone were already well established, but the creativity we newly identify in the fabrication of these artifacts seems certain to reflect the new (or newly enhanced) applications that had been found for them. Fashioned (and for a time also partly decorated) with the bow-drill, they are exemplars of the new technology of the wheel that is more or less simultaneously manifested in pictographic representations of wheeled vehicles and in wheel-made pottery.

The remarkable, and quite unparalleled, spurt of creativity apparent in their iconography can be understood only as a contribution largely to be credited to the artisans producing these objects. It is not a quality that could be infused into their manufacture merely by the issuance of appropriate commands from repressive leaders. Instead we must see it as reflecting in no small degree the artisans' own sense that they were significant participants in larger, ongoing processes of social transformation—the understanding that their innovative contributions would be not merely tolerated but warmly welcomed and long remembered.

Students of the ancient Orient do not ordinarily allow themselves to speak of motivations applying at the level of individuals unless there is what at least purports to be their subjects' own written testimony. But for technology, virtually absent in the written record, this caution imposes unacceptable obstacles to full historical interpretation. The irregularity of waves of innovation will never be understood unless technical and stylistic innovators are seen as active agents in their own right, with substantial areas of discretion and incentive structures of their own.

Paralleling the broad social admiration and support that must have encouraged creativity among seal cutters are trends in other technological sectors. An examination of jewelry and other metal objects from the Royal Tombs of Ur, a few centuries later, has revealed "knowledge of virtually every type of metallurgical phenomenon except the hardening of steel that was exploited by technologists up to the end of the nineteenth century A.D." (Smith, 1972:499). Although this was a society in which principles of social hierarchy were being radically extended and highlighted, it provided, in other words, an incentive structure for the crafts that strongly encouraged creativity. Quite possibly, artisans were allowed to share in some aspects of the enhanced symbolic status of elites in the society. At the same time, it is likely that intensified stratification was for at least a time associated with preponderantly religious rather than secular overtones (Helms, 1993).

There were comparably far-reaching developments in other crafts, reflecting most of the technical and stylistic patterns that prevailed for centuries or even millennia afterward. Great advances were quickly required, for example, in the planning and construction techniques required for public buildings of unprecedented size. Although no direct evidence of woolen textiles has survived, it would be surprising if the same did not apply to the production of textiles, probably beginning to employ fairly large numbers of dependent women or slaves on a quasi-industrial basis even at this early date. The importation already in the Uruk period of substantial quantities of copper indirectly affirms the availability by that time of locally produced commodities like textiles that were valuable, easily transportable, and potentially exchangeable for this needed resource.

Woodworking was surely also a field of important advance, with elite and ritual furniture making entailing further refinements in this field. The same applies to boat construction, which was assuming new importance with the growth of long-distance trade with regions as distant as the Indus Valley. So also, with the introduction of wheeled carts, were simple forms of rotary motion adapted for protracted, heavy-duty use (Nissen, 1991).

In sum, a number of newly specialized technologies including writing advanced at unprecedented rates of innovation. This reaffirms the multiplicity of forces that were at work and the complex, dynamic, "path-dependent" character of the process by which new solutions were found for preexisting sociotechnical problems. Characteristically, the problems initially addressed and solved often did not begin to exhaust the full potential of their original solutions. Cylinder seals, the wheel, and writing are suggestive of a general process in which dawning awareness of an expanding range of uses for new discoveries is likely to have enhanced the creativity lavished on them. This would then lead to initially unforeseeable positive feedbacks, further hastening the overall pace of change.

All these synergistic developments call attention to a richer and more variegated set of processes of change than those centering on the powers and activities of politico-religious elites that have tended to be the almost exclusive object of our attention. Discussants of an important recent set of articles focusing rather narrowly on elite pursuit of political power in prehispanic Mesoamerica have incisively made the same point:

> Studies of ancient symbolic and political systems and human agency tend to take a "top-down" view of power. There are good reasons for this .... Subordinates are much less likely to leave behind recognizable traces of their political strategies and worldviews both because they have fewer means to do so than rulers and because the latter may actively discourage such expressions. There is an increasing body of evidence, mostly from sociology, cultural anthropology, and history, suggesting that, just as elites have their goals, so do those they seek to dominate. Neither are completely successful in their efforts . . . , and even the humblest social unit has some resources with which it can obstruct the actions of the most powerful. . . . Such actions, in turn, require countermeasures, resulting in a dynamic system that refuses to be subsumed by a typological label. (Schortman et al., 1996:62)

Technological change and innovation, I argue, offer important vantage points from which to observe multilevel interactive processes that are not narrowly elite centered. And in so doing, they direct attention beyond the development or imposition of state-level institutions as the defining characteristics of early civilized societies. Childe's original insight, to identify a closely related set of achievements as "urban"—leaving open whether we see the city as some sort of primary stimulus or merely as a container of many new forms of innovation and interaction at many levels—in this sense still retains at least some of its attractiveness as an alternative to formulations identifying the emerence of Mesopotamian civilization primarily with the crystallization of state power.

DYNAMICS OF THE TRANSITION FROM BRONZE TO IRON

A second major example of an accelerated transition having an important technological as well as social character involves the substantial replacement of bronze by iron beginning late in the second millennium B.C. Profoundly important as the introduction of iron metallurgy presently became for all the Near Eastern societies and economies of the time, it unfortunately occurred almost entirely outside the range of ancient scribal—and hence elite—attention. Before turning to the specifically metallurgical aspects of the transition, however, we can obtain important insights into them by considering a concurrent feature of pyrotechnological innovation that happens not to have suffered as much from scribal neglect.

The manufacture of glass is the outstanding exception to the prevailing absence of scribal coverage of craft procedures. Techniques for producing thin glazes and frits had long been known. Now they were abruptly supplemented by the introduction of highly prized, multicolored glass vessels. Cuneiform texts of El Amarna age (c. 1350–1334 B.C.) attest to a new and almost insatiable appetite for glass on the part of Near Eastern rulers and their courts who were the prime consumers of luxuries. Preserved in the Assurbanipal Library, if in later copies that describe procedures in a complex literary tradition well removed from actual technological practice, are instructions for its manufacture.

Leo Oppenheim, in one of the last contributions reflecting his astonishingly varied range of interests, has persuasively accounted for this exception. From a probable source of discovery in upper Mesopotamia, Mitanni craftsmen in the 15th and early 14th centuries B.C. suddenly introduced alkali-silicate glass of high technological and stylistic sophistication all over the Near East. As Oppenheim wrote,

> What actually happened was a change in glass technology. Instead of producing glazes that were used like paint on formed objects, craftsmen now created opaque and intensely colored glasses; they fused and fashioned them into objects made entirely of glass. Unprecedented techniques were applied to shape the glass when still hot and this resulted in the beautifully decorated vessels and flasks of that period. (1973:262)

Very little is known of the circumstances in which the invention of glass originated, for we see it only in its patterns of diffusion and consumption. Some glass circulated as items of trade and royal exchange, but much of it was apparently made to order by craftsmen who had been persuaded (or more likely compelled) to come from the source region to the courts of other princes to carry on their work. It is noteworthy, however, that the art quickly stagnated in those more distant centers; the acquisitive whims of royal sponsorship apparently did more to stifle than to encourage it. In its upper Syrian homeland, on the other hand, glassmakers went on to learn how to make translucent glass and how to use a blowpipe and molds to produce more complex forms.

The Assurbanipal Library texts go back to 12th-century or earlier originals. Also found there, and of exactly the same genre, are prescriptions for making perfumes and for training fast chariot horses. What all these represent, Oppenheim suggested, is

> a concerted effort of the Assyrian royal court to promote crafts considered at that time essential for royal prestige. Apparently every ruler of the Amarna Age and the subsequent centuries simply had to have fast chariot horses, new perfumes and beautiful glassware. . . . The desire of the Assyrian court to become independent of this interchange of information carried typically by migrating craftsmen and technicians may have prompted the writing down of the essential of these three crafts. (1973:265)

The apparently complete detachment of the interests of these rulers from direct support of pyrotechnological crafts must be emphasized, for it finds an exact counterpart in their attitudes toward iron metallurgy. Royal involvement was wholly noneconomic—even antieconomic—in character and was instead confined to acquiring and monopolizing prestige goods like ornamental glass, perfumes, and fast horses. Prized objects of glass were suitable gifts in the courtly exchanges that characterized the relations between "great kings" in the El Amarna period. Like physicians and soothsayers, craftmen conversant with their manufacture also were no doubt involuntary objects of loan and barter between their royal masters (Zaccagnini, 1990). But this pattern of utilization was wholly inconsistent with the dissemination of knowledge of the technique and its products into widespread, popular use.

To return now to iron metallurgy, smelted iron had been known and used in very small quantities for nearly two millennia. As its long-continuing rarity and use as little more than an exotic substance make clear, iron's potentially valuable properties were slow to be reliably reproduced or perhaps even recognized. Hittite texts of the 16th century B.C. are the first to refer to considerable numbers of utilitarian objects made of iron. Some, including a throne given to Anitta of Kanesh, were apparently of considerable size and complexity. At that early time, however, the production of smelted iron still seems to have been only irregularly attainable.

More important in any case were two further steps—carburization and tempering—required for the production of steel. A working knowledge of both was required before iron as a basic material could be modified to regularly exceed bronze in either tensile strength or hardness. Neither in mastering these processes nor in the original discovery of iron itself can the special role sometimes claimed for Anatolia be substantiated.

Iron ores are vastly more plentiful than those of copper and tin. Copper and iron ores are often found together in nature, however, and iron is a common impurity in early bronze artifacts in the Near East. Early Cypriot copper slag has, in fact, been mined as a source of iron ore in relatively recent times. Thus it is very likely that knowledge of iron working began in centers already specialized in copper working.

Appreciably higher temperatures are required than for smelting copper. Hence significant growth in the scale of iron production was probably accompanied by the introduction of somewhat more complex smelting techniques that included the use of bellows to produce forced-air drafts. Iron production also entails more arduous hand labor, both to concentrate the ore and to separate the molten metal from the slag in the spongy bloom by prolonged hammering.

For these reasons, compounded by its less attractive appearance and susceptibility to rust, iron remained of very limited use until well after the conventional ending of the Bronze Age. Weapons of iron first began to appear occasionally in Syro-Palestinian tombs in the 12th century B.C., and iron agricultural tools only in the following century. Not before the 10th century (and somewhat later in Mesopotamia) did iron apparently become the predominant material for both. In Babylon of Hammurabi's time (c. 1792–1750 B.C.), iron was still reckoned eight times as valuable as silver. By the mid-first millennium, it had become cheaper than copper, and silver was reckoned from 300 to 600 times costlier than iron. Well before that time, by no later than the eighth century B.C., putting citizenry to the iron dagger had become a metaphor for great slaughter by Assyrian armies, and iron fetters figuratively asserted the power of Assyrian kings (Curtis et al., 1979; Muhly et al., 1982; Muhly et al, 1985; Pleiner and Bjorkman, 1974; Waldbaum, 1978; Wertime and Muhly, 1980; Yahya, 1995).

There was, in short, a relatively rapid transition to the full exploitation of iron after a very long interval marked by almost no progress in this direction. This abrupt increase in the rate of sociotechnical change parallels the broader sociotechnical changes accompanying the urban revolution that were mentioned earlier. In the earlier instance, I have argued that dramatic, ongoing changes in social scale and complexity supplied the necessary setting for this to be understood. Do comparable contextual features also in this case that help to account for the impulsive irregularity in rate of change?

Profound socioeconomic changes were certainly underway in the late second millennium B.C., although at first glance they are puzzlingly different in character from the earlier ones. Focusing in particular on Syria and the Levant, Liverani (1987) described the end of the Late Bronze Age as a time of crisis. Precipitating the crisis, he suggested, was an increasing concentration of organizational authority and conspicuous consumption of luxury goods in palaces of the period. That led to "a crude exploitation of the village communities" and the consequent "degradation of basic, domestic culture." Informal, intrafamily or -village responses to differential needs were eroded, for example, "by the growing consideration of land and work as merchandise." The traditional, periodic cancellation of debts became more irregular or disappeared. Hence runaways increased in number; settlements shrunk back toward the better-watered areas; nomads and seminomadic groups became more prominent; and rural insecurity increased correspondingly. As "great kings" disappeared from the scene, the crisis extended to "inter-regional contacts requiring their re-casting in another form." So it was that the collapse of palaces precipitated more pervasive disasters affecting entire kingdoms (Liverani, 1987:69–70).

In Liverani's view, this was an essentially devolutionary process; clearly this was the case in many respects. The scale of at least the larger political configurations declined abruptly. Long-distance, centrally organized trade in which the merchant of the Late Bronze Age "was active for his own profit only as a sideline" gave way to more localized Iron Age patterns in which "the very supply of raw materials becomes an activity which is carried out for profit, stimulated not by royal order but by advantage."

> For the large monopolies of metal production of the LBA, connected by efficient palace trade, the Iron I period substitutes metal production on a local scale. The same holds true for metal working: the rather sophisticated palace installations necessary for bronze working are substituted by the iron worker's much more accessible and movable equipment. (Liverani, 1987:71–72)

Still, the question of cui bono cannot be avoided. On whose behalf were the greater scale and efficiency of Late Bronze Age trading patterns exercised? If the bronze-working installations of the palaces were indeed relatively more sophisticated than those that followed, of what benefit was this to the greater part of the population whose well-being was concurrently being eroded? What pattern of inducements did the Late Bronze Age monarchies offer to artisans to stimulate their creativity?

I suggest that accompanying the technological changes of the Iron Age was a weakening of autocratic control and a consequent widening of access to craft-produced goods. Supplies of iron for popular consumption multiplied many times, and its cost fell dramatically, even as the technological superiority of iron and steel for tools and weapons advanced enormously and became widely known. Although some of these new uses had bronze prototypes, others like saws, dies, shears, tweezers, and lathes were either new or massively modified to take advantage of the unprecedented properties of the new raw material. Here, too, we obtain a glimpse of a reciprocal process. New material capabilities stimulated new designs, and these in turn hastened further improvements in the materials.

The rapidly growing set of new iron tools and weapons soon began to benefit agriculture, construction, and virtually every productive sector. Some reached the larger public by way of distributions of captured loot or through approved sales by palace craftsmen. If my surmises on the radical dispersion of the new industry are correct, however, nonroyal workshops producing directly for the public played an increasingly important part. This, I submit, was the favorable context that finally liberated the iron-working industry for its impulsive advance into what is deservedly given the designation of being a new age. The fragmentation of political power, or more accurately the freedom of action that could flourish in the interstices of a continually fragmenting pattern, created favorable conditions for technological innovation and specialization.

As this occurred, the ties must have loosened between newly specialized blacksmiths now turning their attention exclusively to iron and the copper-mining and -smelting communities and workshops where they had originated. Separate specialties often began to be listed even in palace accounts, where the role of a unified command economy in amassing supplies of raw materials would tend to keep them together. Elsewhere, specialization probably proceeded faster and further.

Recoverable iron ore deposits are more dispersed than Near Eastern sources of copper, so that supply networks became much denser and more difficult for palaces to control and monopolize. Iron ore was so plentiful that most of its sources remain unknown to us. Emergent shortages of wood for charcoal in the vicinity of traditional smelting locations were probably an additional factor promoting dispersal. It is generally assumed that contemporary iron-ore reduction methods were several times as fuel efficient as those for copper.

Similarly, we should expect a secondary (but economically most significant) transformation of productive capabilities in fields like land clearance for agriculture, transportation, and building. I am not aware that the vast improvements that iron permitted in wheeled transportation have received much attention, but as the well-known biblical passage attests, they surely deserve it:

> And the Lord was with Judah; and he drove out the inhabitants of the mountain; but could not drive out the [Canaanite] inhabitants of the valley, because they had chariots of iron. (Judges 1:19)

In this case as in others, we suffer from the attachment of many contemporary scholars, especially (but not only) those linked to museums, to recoverable objects that have often been removed from their archaeological context. If the aim is to understand better the nature of technological change, we need to widen our focus from *products* to include greater emphasis on the *processes, organizations, and human agents* by which they were made (Stein, 1998:18–21). In my earlier example of the Urban Revolution, that applies to the character of administrative processes. In this case, it suggests that we need to focus more attention on advancing mastery of pyrotechnological processes, rather than simply taking for granted an economically driven desire for iron tools and weapons (cf., e.g., Wertime and Wertime, 1982).

There are questions here that are no less applicable to other, more modern historical periods and settings. Does a demand for new products regularly induce innovative methods for producing them, or does the discovery of new manufacturing processes liberate designers to create new products presupposing their use? How much does the nature of the demand matter—royal whim or market, for example? We may never be able to follow that question to a definitive conclusion in the ancient Near East (and often cannot do so even in the modern world), but it is surely safest to assume an interactive, iterative, evolutionary process in which both supply and demand continuously influence each other.

Once again, it is clear that archaeologists need consistently to maintain a widened span of attention, interacting more closely with historians and embracing sociotechnical contexts and processes as well as products. If it is to offer insights for the larger subject of disjunctive change in history, a focus on technological change should not be thought of as merely leading to an enumeration of growing lists of inanimate, material goods. Instead we must seek to identify the agents of change and the character of the incentives that drove them. Moreover, our field of concern needs to include the impact of technological changes on structures of distribution and patterns of demand—and the reciprocal influences of the latter on the ingenuity and diligence with which technological innovation was itself pursued.

TECHNOLOGY AND THE DECLINE OF THE ROMAN EMPIRE

My third example, venturing well outside those regions with whose primary archaeological and historical resources I can claim personal familiarity, can be put forward only briefly and tentatively. It involves taking a skeptical, technology-centered view of the stagnation and retrogression that is usually attributed to the Roman Empire before its collapse in the west in the fifth century A.D.—an evaluation that retains many similarities to that of Gibbon's magesterial *Decline and Fall of the Roman Empire* written more than two centuries ago.

Historical generalizations applying with little differentiation to late Roman society as a whole have until recently tended to be drawn, just as they were in Gibbon's day, primarily from normative literary, upper class or aristocratic sources. It was manifestly a society that to a considerable extent depended for its leisure and wealth on unremunerated labor and passive ownership of great agricultural estates. Hence it is hardly surprising that individuals near the apex of society, writing of their own proclivities and for their peers, expressed unqualified repugnance for active roles in agricultural management, or indeed for any form of gainful employment, and for all those having to engage in it. "In general, they were held in bondage by a value system that emphasized consumption rather than productive investment" (Garnsey and Saller, 1987:74).

The almost inevitable interpretation to be drawn from these and many other sources—viewing the economy and its supporting technology through the special prism of its urban-based, political and literary elites—is that late Roman society (not to speak, for present purposes, of Graeco-Roman civilization as a whole!) was fundamentally deficient in economic enterprise and consequently in technological innovation as well. Moses Finley (1973), the most prominent recent commentator on the subject, characterized the economy in this way, as fundamentally agrarian and tribute based. He buttressed this view with a summary of known Graeco-Roman technological contributions, underlining their apparent paucity. Manufacturing was not called on to provide a basis for interregional exchange, furthermore, because cities could be centered almost exclusively on consumption—essentially all their requirements being met instead through the exaction of taxes and rent. All sectors of activity during the Roman Empire are thus said to have been characterized by relative technological stagnation. Increased agricultural productivity and output, insofar as they may have occurred, are attributed mainly to extensions of cultivated areas and increased labor inputs rather than to technological improvements (Finley, 1965:29, 32; 1973:75–76; Garnsey and Seller, 1987:57, 63, 198; Pleket, 1967:19; cf. Adams, 1996:44–45).

Gathering force and cohesiveness in recent years, however, is a view that, while multifaceted and heterogeneous, is substantially contrary. Even though relying in part on the same kinds of primarily literary, epigraphic, and historiographic sources, some authorities have begun to find grounds for identifying a much more diverse set of attitudes and patterns of behavior even on the part of the Roman elite (Frederiksen, 1975). In addition, there have been innovative applications of social science analytical methodologies to reports on the performance of the Roman political economy. Extending across the spectrum, from econometric history to Marxism, these revisionary contributions obviously diverge sharply from one another in many respects. Taken together, however, they support the case for more economically effective land management on the part of owners of large Roman estates than the existing literary statements suggest (e.g., Gunderson, 1976; Wickham, 1984).

In many ways, however, the most substantial source of revisionist criticism is archaeological. For the most part, it has arisen not in studies of the major imperial centers but in outlying provinces on the imperial peripheries—in Syro-Palestine on the east (cf. Kennedy et al., 1993) and especially in Britain and Gaul.

One theme of the rising debate in the west involves the employment of water wheels for milling grain, an invention that is first attested in the early Imperial period. As "the first applied use of inorganic power in economic production," its strategic importance should have been clear from the outset (Anderson, 1974:80). Yet the prevailing interpretation, based on the paucity of literary references, is that water wheels were not significantly exploited until the later Middle Ages.

Scholars with a primarily archaeological background are naturally disposed to skepticism where the evidence is largely of this character. If the stereotypes of absentee management have any validity, many labor-saving practices might be introduced in the countryside of which urban-based landowners (and those writing for them and on their behalf) remained entirely ignorant. Hence archaeologically inspired efforts have begun to controvert the impression of water-wheel scarcity. Although the contrary evidence assembled to date is still modest, it has even been suggested that, at least in Gaul and Britain, water wheels may have been part of the *normal* equipment of milling establishments during the later Roman Empire (Greene, 1990:215; Raepsaet, 1995:916; Wikander, 1984:7, 9–10, 22).

Increasingly outspoken dissatisfaction on the part of archaeologists is being expressed over what is seen as the stale reiteration and lack of research progress that more traditional approaches have led to. Following the example of prehistorians, their alternative approach depends primarily on new knowledge deriving from analytical techniques developed by natural scientists and from stratigraphic excavation of ordinary living refuse rather than literary references. At the same time, scenes depicting complex agricultural implements on Roman gravestones are among other, previously little-used source of evidence. As they accumulate, these new leads seem to call for a substantially more innovation-based reconstruction of agricultural and craft technology than had previously existed.

Only greater archaeological self-reliance, it is asserted, can correct the "clear bias" reflected in

> the emphasis on "Roman" attributes of [Britain] with a relative disregard of the native. Attention is focused on the rich and wealthy, and little effort is directed to the study of the poor. The quantity of attention lavished on the excavation of villas, walled towns and forts and the relative absence of detailed studies of non-villa settlements and unwalled small towns demonstrates the bias. . . . The approach does not give a true picture of Romano-British society or of the history of the province. (Hingley, 1989:3)

The rhetoric accompanying some examples of this approach is described by a recent critic as "barbarian by conviction, willingly rejecting classical wisdom." It is partly open to rebuttal that increasing but unspoken dependence is implied

> on ancient historians for their knowledge of the tax systems, social structures, and religious beliefs of the conquerors. References to classical texts continue to appear, but in token (and unusable) forms (rather like the continued appearance of eagles on barbarian jewelry even after their original significance and function was no longer fully understood). Ripped out of

> their contexts, with little concern for the processes of formation that resulted in our complex literary deposits, odd snippets of classical authors are treated with a brutal insensitivity to which the new Romanists would not subject the humblest fragment of animal bone or iron slag. (Woolf, 1996:8)

But as this critic went on to concede, "The results have more than justified the polemics." Britain's status as the best excavated, most comprehensively surveyed region of the Roman Empire is secure. Roman archaeology in Britain, in particular, has become an important laboratory for bringing together historical and archaeological approaches more generally. And there at least, not only the cultural diversity characterizing everyday life in the smaller towns and countryside, but also the tenuousness of the ties of this outlying, increasingly "Un-Roman-looking" province to the Roman Empire have been highlighted (Greene, 1986, 1990; Lewit, 1991; Millett, 1990; Raepsaet, 1995).

It would be foolhardy for a non-Romanist to venture a guess as to how far the traditional picture will presently be altered. But at the very least, there is strong reinforcement here for a more continuous, better institutionalized dialogue between archaeologists and historians on a wide range of economic, technological, demographic, and other issues.

TECHNOLOGY AND THE COMPLEMENTARITY OF APPROACHES

The complementarity of archaeological and historical sources, wherever both exist for the same civilizational entity, is surely not a matter of dispute. Individuals once tended to receive training only in one or the other discipline, but that narrowness of specialization has now been widely recognized as inimical to the natural end of interdisciplinary understanding. What still remains insufficiently recognized, however, is the full range of ways in which textual and archaeological evidence should be exploited to maximize the level of understanding to which both can contribute.

This discussion focusing on ancient technological change, defined broadly enough to take account of its cultural and socioeconomic contexts and to include its organizational dimensions, is intended to clarify some of the major meanings of a stance of complementarity for archaeologists in particular. What such a stance should *not* involve is a major, let alone exclusive, preoccupation with the same ancient traditions, institutions, and activities as are disclosed by textual sources, embodying the same biases and limitations as those sources. Archaeological and historical approaches should naturally be brought into effective convergence, but complementarity also means that each must recognize and exploit its own inherent potentialities as a contribution to the joint effort.

Briefly to summarize, consideration of the role of technology in the beginnings of urban civilization in Mesopotamia, the onset of the Iron Age in the larger Near East, and the Roman Empire in the west before to its collapse in the fifth century A.D. has led to an identification of a number of distinctive contributions that archaeology can make to the common enterprise. Some are already widely practiced and well understood and have not been dealt with here. These include, for example, holistic descriptions of everyday life and the identifications of secular changes in settlement, economy, and demography. But an attempt has been made to show that others like the following deserve more attention than they are currently receiving.

1. Local and regional, as well as urban and rural, differences in wealth, cultural inventories, subsistence patterns, ethnic composition, access to trade and market networks, and so on, are of fundamental importance as themes of archaeological investigation. Typically overlooked or treated only impressionistically in textual sources, they have much to tell us about the uniformity, coherence, and dynamics of change in large-scale territorial systems and ancient empires.
2. No less central as a focus of concern for which archaeological evidence is of primary importance are secular changes in the broad class or status composition of society. Textual evidence may or may not also be available, but the possibility must not be overlooked that familiar archaeological indices like housing, household furnishings, burial contents, and skeletal stature provide independent windows of observation that support different interpretations. Changing ranges of variability in these indices rather than merely what are asserted to be "typical" examples should be the key themes of observation. The objective is to develop comparisons and contrasts with the skewed observations on these subjects that are the most that can be expected from ancient textual sources.
3. Continuities or disruptions in daily, local life need to be compared and contrasted with their counterparts in traditional historical sources that reflect "centers" rather than "peripheries." Herein are likely to be found clues to systemic disorders and breakdowns.
4. Similarly, comparisons and contrasts bearing on sources of innovation and initiatives for change should be sought. Those that can be derived from independent, broadly based archaeological inquiry can serve to complement and correct the bias that is naturally found in either archaeological monuments or textual records focusing almost exclusively on elite ideologies and activities.
5. Consistent with the broadly inclusive approach to technology adopted here, there should be a broadening and shift in the focus of archaeological studies of technology more generally. It is important for the latter regularly to extend beyond artifacts or *products* to take more fully into

account physical *processes* of manufacture, organizations engaged in resource acquisition, production, and distribution, and—insofar as they can be identified—the human agents engaged in these activities.

What technology constitutes, in other words, is a theme that forces the integration of archaeological and historical approaches into an integrative framework that accomodates the distinctive contributions that each can make and can help to circumvent some of their respective limitations. Systems of production, distribution, and consumption, to be sure, were never autonomous. In varying degrees, they remained subject to hierarchical control by politico-religious elites as well as by the needs and demands of even broader social forces. But they also, and even more directly, implicate the accomplishments, traditions, and associations of a surely strategic set of actors about whom our textual sources, in particular, are generally silent. An intensified archaeological focus on technology affords a very real prospect of exposing patterns of structural diversity, conflicts of interest, and widely distributed centers of initiative. All these matters are central to an understanding of the dynamics of change, of which we are unlikely to learn much in any other way.

Technology, if it is to serve this role, overrides any spurious distinction between "mere" skill and empowering intellect. What is called for is in one sense a highly synthetic approach, drawing together different realms of activity and data. At the same time, there is also a sense in which this approach is deconstructive, consistently seeking to penetrate behind the formal, holistic representations found in politico-religious streams of tradition and their protagonists. Yet this is not the expression of a narrowly materialist view of history. The essential, universal elements it embodies include the organization and communication of decision making, the modification and reproduction of human capabilities, and the vital human process of groping for and asserting certainties while condemned forever to navigate in a stormy sea of uncertainties.

ACKNOWLEDGMENTS

Taking a different direction, this chapter draws on themes briefly discussed in Adams, 1996. Its subject emerged only gradually, beginning with the presentation of earlier versions at the Istituto di Paletnologia of the University of Rome and the Seminar für Vorderasiatische Altertumskunde of the Free University of Berlin in May and June 1996. I am grateful to discussants on both of those occasions, and to Guillermo Algaze, Thomas E. Levy, and Hans J. Nissen, for critical comments. Finally, I am most grateful to the Wissenschaftskolleg zu Berlin for many forms of generous support during the initial stage of this chapter's preparation.

REFERENCES

Adams, R. McC., 1996, *Paths of Fire: An Anthropologist's Inquiry into Western Technology*. Princeton University Press, Princeton.

Adams, R. McC., in press, Complexity in Early State Societies. In *A Companion to Archaeology*, edited by J. Bintliff. Blackwell Publishers, Oxford.

Anderson, P., 1974, *Passages from Antiquity to Feudalism*. NLB, London.

Brooks, H., 1980, Technology, Evolution, and Purpose. *Daedalus* 109(1):65–81.

Childe, V., 1950, The Urban Revolution. *Town Planning Review* 21:3–17.

Curtis, J. E., Wheeler, T. S., Muhly, J. D., and Maddin, R., 1979, Neo-Assyrian Ironworking Technology. *Proceedings of the American Philosophical Society* 123:359–390.

Damgaard, C., 1996, The Rate of Evolution in Growing Populations. *Trends in Ecology and Evolution* 11:107–108.

Eldredge, N., and Gould, S. J. 1972, Punctuated Equilibria: An Alternative to Phyletic Gradualism. in *Models in Paleobiology*, edited by T. J. M. Schopf, pp. 305–332. Freeman, Cooper, and Co., San Francisco.

Finley, M. I., 1965, Technical Innovation and Economic Progress in the Ancient World. *Economic History Review* 17:29–45.

Finley, M. I., 1973, *The Ancient Economy*. University of California Press, Berkeley and Los Angeles.

Fitch, W. M., and Ayala, F. J. (editors), 1995, *Tempo and Mode in Evolution: Genetics and Paleontology 50 Years after Simpson*. National Academy Press, Washington, D.C.

Frankfort, H., 1939, *Cylinder Seals*. Macmillan, London.

Frederiksen, M. W., 1975, Theory, Evidence, and the Ancient Economy. *Journal of Roman Studies* 65:164–171.

Garnsey, P., and Saller, R., 1987, *The Roman Empire: Economy, Society and Culture*. Duckworth, London.

Girifalco, L. A., 1991, *Dynamics of Technological Change*. Van Nostrand Reinhold, London.

Gould, S. J., and Eldredge, N., 1993, Punctuate Equilibrium Comes of Age. *Nature* 366:223-227.

Greene, K., 1986, *The Archaeology of the Roman Economy*. B.Y. Batsford, London.

Greene, K., 1990, Perspectives on Roman Technology. *Oxford Journal of Archaeology* 9:209–221.

Gunderson, G., 1976, Economic Change and the Demise of the Roman Empire. *Explorations in Economic History* 13:43–68.

Helms, M. W., 1993, *Crafts and the Kingly Ideal: Art, Trade, and Power*. University of Texas Press, Austin.

Hingley, R., 1989, *Rural Settlement in Roman Britain*. Seaby, London.

Kennedy, D., Isaac, B., and Millar, F., 1993, *The Roman Near East, 31 B.C.–A.D. 337*. Harvard University Press, Cambridge.

Lewit, T., 1991, *Agricultural Production in the Roman Economy A.D. 200–400*. BAR International Series 568. British Archaeological Reports, Oxford.

Liverani, M., 1987, The Collapse of the Near Eastern Regional System at the End of the Bronze Age. In *Centre and Periphery in the Ancient World*, edited by M. Rowlands, M. Larsen, and K. Kristiansen, pp. 66–73. Cambridge Univerity Press, Cambridge.

Marcus, J., and Flannery, K.V., 1996, *Zapotec Civilization: How Urban Society Evolved in Mexico's Oaxaca Valley*. Thames and Hudson, New York.

Mayr, E., 1988, Speciational Evolution through Punctuated Equilibria. In *Toward a New Philosophy of Biology: Observations of an Evolutionist*, edited by E. Mayr, pp. 457–488. Belknap Press of Harvard University Press.

Millett, M., 1990, *The Romanization of Britain: An Essay in Archaeological Interpretation*. Cambridge University Press, Cambridge.

Muhly, J. D., Maddin, R., Stech, T., and Özgen, E., 1985, Iron in Anatolia and the Nature of the Hittite Iron Industry. *Anatolian Studies* 35:67–84.

Muhly, J. D., Maddin, R., and Karageorghis, K. (editors), 1982, *Early Metallurgy in Cyprus, 4000–500 B.C.* Pierides Foundation, Nicosia.

Nissen, H. J., 1991, Die gelagerte Achse. Paper presented at Colloquium on Handwerk und Technologie im Alten Orient, Berlin.

Nissen, H. J., Damerow, P., and Englund, R. K., 1993, *Archaic Bookkeeping: Early Writing and Techniques of Economic Administration in the Ancient Near East.* University of Chicago Press, Chicago.

Oppenheim, A. L., 1973, Toward a History of Glass in the Ancient Near East. *Journal of the American Oriental Society* 93:259–266.

Pleiner, R., and Bjorkman, J. K., 1974, The Assyrian Iron Age: The History of Iron in the Assyrian Civilization. *Proceedings of the American Philosophical Society* 118:283–313.

Pleket, H., 1967, Technology and Society in the Graeco-Roman World. *Acta Historica Nederlandica* 2:1–25.

Postgate, N., 1992., *Early Mesopotamia: Society and Economy at the Dawn of History.* Routledge, London.

Potts, R., 1996, Evolution and Climate Variability. *Science* 273:922–923.

Raepsaet, G., 1995, Les prémices de la méchanisation agricol entre Seine et Rhin de l'antiquité au 13e siècle. *Annales HSS* July–August:911–942.

Rothman, M. S. (editor), in press, *Mesopotamia in the Era of State Formation.* School of American Research Press, Santa Fe.

Schortman, E., Urban, P., and Ausec, M., 1996, Comment on R.E. Blanton et al., A Dual-Processual Theory for the Evolution of Mesoamerican Civilization. *Current Anthropology* 37:61–63.

Schumpeter, J., 1939, *Business Cycles: A Theoretical, Historical, and Statistical Analysis of the Capitalist Process.* McGraw-Hill University Press, London.

Smith, C. S., 1972, Technology and Science: Notes on their Historic Interaction. *Technology and Culture* 11:493–549.

Spencer, C. S., 1990, On the Tempo and Mode of State Formation: Neoevolutionism Reconsidered. *Journal of Anthropological Archaeology* 9:1–30.

Stein, G. J., 1996, Producers, Patrons, and Prestige: Craft Specialists and Emergent Elites in Mesopotamia from 5500–3100 B.C. In *Craft Specialization and Social Evolution: In Memory of V. Gordon Childe*, edited by B. Wailes, pp. 25–38. University Museum Monograph No. 93. University of Pennsylvania, Philadelphia.

Stein, G. J., 1998, Hetrerogeneity, Power, and Political Economy: Some Current Research Issues in the Archaeology of Old World Complex Societies. *Journal of Archaeological Research* 6:1–44.

Stein, G. J., and Blackman, M. J., 1993, The Organizational Context of Specialized Craft Production in Early Mesopotamian States. *Research in Economic Anthropology* 14:29–59.

Waldbaum, J., 1978, *From Bronze to Iron: The Transition from the Bronze Age to the Iron Age in the Eastern Mediterranean.* Humanities Press, Atlantic Highlands.

Wertime, T. A., and Muhly, J. D. (editors), 1980, *The Coming of the Age of Iron.* Yale University Press, New Haven.

Wertime, T. A., and Wertime, S. F. (editors), 1982, *Early Pyrotechnology: The Evolution of the First Fire-Using Industries.* Smithsonian Institution Press, Washington, D.C.

Wickham, C., 1984, The Other Transition: From the Ancient World to Feudalism. *Past and Present* 103:3–36.

Wikander, Ö., 1984, *Exploitation of Water-Power or Technological Stagnation? A Reappraisal of the Productive Forces in the Roman Empire.* Scripta Minora Regiae Societatis Humaniorum Litterarum Lundensis 1983–1984. CWK Gleerup, Lund.

Woolf, G., 1996, Ruled Britannia. *Times Literary Supplement* April 12:8.

Yahya, A. H., 1995, *The Coming of the Iron Age in Palestine.* Unpublished Ph.D. dissertation, Fachbereich Altertumswissenschaften, Free University of Berlin, Berlin.

Zaccagnini, C., 1990, The Transition from Bronze to Iron in the Near East and in the Levant: Marginal Notes. *Journal of the American Oriental Society* 110:493–502.

Zettler, R. L., 1996, Gordon Childe and the Socioeconomic Position of Craft Specialists in Early Mesopotamia. In *Craft Specialization and Social Evolution: In Memory of V. Gordon Childe,* edited by B.Wailes, pp. 17–23. University Museum Monograph No. 93. University of Pennsylvania, Philadelphia.

Chapter 5

High-Intensity Household-Scale Production in Ancient Mesoamerica

A Perspective from Ejutla, Oaxaca

GARY M. FEINMAN AND LINDA M. NICHOLAS

INTRODUCTION

For two decades, archaeological perspectives on nonagricultural economic specialization (e.g., Peacock, 1982; Santley et al., 1989; van der Leeuw, 1976, 1977) have tended to categorize variability in a monolithic manner, viewing craft specialization along a single dimension, from small-scale household production to larger-scale factory manufacture (Figure 1). Yet acceptance of this view has created conundrums at the Mesoamerican sites of Colha and Teotihuacan (Figure 2), where high densities of stone tool manufacture have been noted, but nonresidential workshops or "factories" have not been found.

This chapter builds on Cathy Costin's (1991) alternative, multidimensional framework for categorizing craft specialization. We present archaeological findings from an excavated Classic-period (A.D. 200–800) house in highland Ejutla, Oaxaca, Mexico (Figure 2), where a heavy volume of craft production appears to have been enacted in a domestic context. Based on the Ejutla findings and a re-

---

GARY M. FEINMAN AND LINDA M. NICHOLAS • Department of Anthropology, The Field Museum, Chicago, Illinois 60605.

Cultural Evolution: Contemporary Viewpoints, edited by Gary M. Feinman and Linda Manzanilla. Kluwer Academic/Plenum Publishers, 2000.

MONOLITHIC MODELS OF CRAFT SPECIALIZATION

VAN DER LEEUW 1977	SANTLEY ET AL. 1989
LARGE SCALE INDUSTRY	MANUFACTORY
	NUCLEATED INDUSTRY
VILLAGE INDUSTRY	
WORKSHOP INDUSTRY	WORKSHOP INDUSTRY
INDIVIDUAL INDUSTRY	
	TETHERED SPECIALIZATION
HOUSEHOLD INDUSTRY	HOUSEHOLD INDUSTRY
HOUSEHOLD PRODUCTION	HOUSEHOLD PRODUCTION

Figure 1. Unilinear models of craft specialization.

consideration of other Mesoamerican contexts, the presumed frequent occurrence of nonresidential craft workshops in ancient Mesoamerica is questioned, although the existence and importance of high-intensity craft manufacture are not. In other words, we suggest that specialized craft production occurred in prehispanic Mesoamerica (and high volumes of nonagricultural goods were produced for exchange), but these activities generally took place in domestic or residential contexts. The implications of this perspective for understanding prehispanic Mesoamerican economies are discussed briefly in a concluding section.

MONOLITHIC MODELS OF CRAFT SPECIALIZATION

According to monolithic models of craft production, household production is defined as small-scale manufacture for one's own residential unit. Household indus-

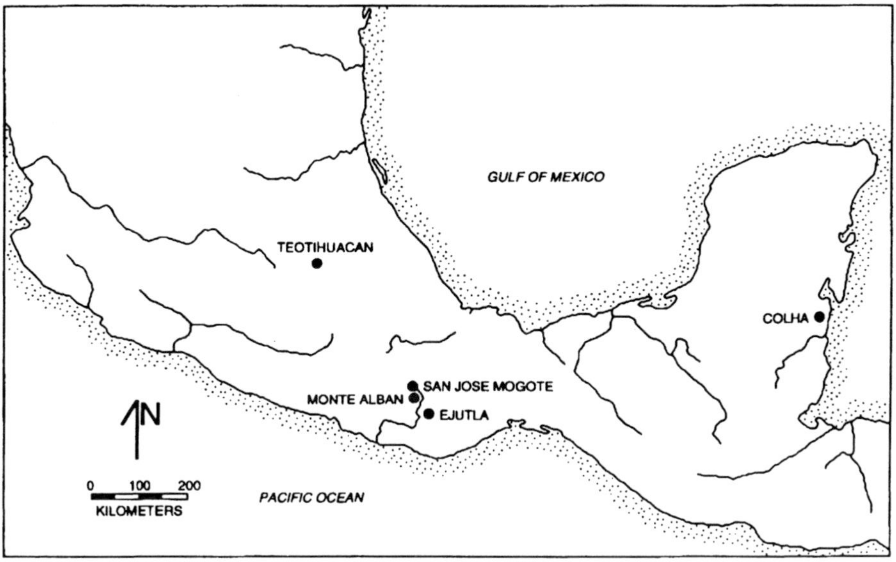

Figure 2 Mesoamerica, showing places mentioned in the text.

try is described as "part-time" production for use by the immediate group, involving entirely local materials and requiring little labor or technological investment. The volume of production is presumed to be low. In contrast, workshop and factory production are viewed as full-time specializations in which increasing degrees of high-intensity production are employed to supply ever wider networks of consumers. Greater degrees of capital investment and more distant raw materials are incorporated into the production process.

These monolithic models expect general agreement among the volume of finished goods that craftworkers produce, the intensity of their work effort, the proportional weight given to craft activities in household income, the density of debris produced, the spatial extent to which finished products are distributed, the distances from which raw materials are procured, the size and elaboration of production facilities, and the scale or setting of the craft activities. Residential contexts are presumed to be associated with part-time, low-intensity, locally focused production, whereas high-intensity, full-time, economically important craftwork is expected in nondomestic settings.

In the following discussion, general acceptance of the monolithic model is shown to have led to interpretive logjams in two well-known cases of prehispanic Mesoamerican craft production. Here, we briefly review these debates and begin to map a path out of this impasse.

SCALE VERSUS INTENSITY: THE STONE TOOL CONUNDRUM

The presence of large and unusual quantities of lithic debris has long been recognized at the central Mexican urban center of Teotihuacan and the Lowland Maya settlement of Colha in northern Belize. Nevertheless, at both sites, parallel debates have raged over the interpretation of this debris. Key tenets of the monolithic model of craft specialization have clouded this debate. Those archaeologists who have argued for large-scale workshop manufacture at Teotihuacan (e.g., Santley, 1983; Spence, 1967, 1981) and Colha (e.g., Hester and Shafer, 1994; Shafer and Hester, 1983, 1986, 1991) have relied principally on massive debris deposits to establish their case for large-scale mass production. At Teotihuacan, high densities of obsidian debris have been noted on the surface of the site; great quantities of local chert have been unearthed during excavations at Colha.

Yet, importantly, no clear nondomestic workshop or factory contexts have been uncovered at either site. In fact, preliminary indications at both sites suggest that craftwork was, indeed, carried out in domestic contexts. For example, at Teotihuacan, Spence (1987:434) noted that many of his dense obsidian concentrations coincide spatially with an architectural unit of some sort, usually a standard Teotihuacan multiroom residential apartment compound. Likewise, at Colha, chert debitage tends to trail off the edge of residential platforms (Roemer, 1982; Shafer and Hester, 1983:529). In excavated contexts, the proximity of craft debris with contempory dwellings seems to indicate that the production activities were undertaken in residential settings (McAnany, 1993:233).

In part, the problem for the interpretations offered by these Teotihuacan and Colha investigators is that they have conflated two different parameters, or dimensions, of craft specialization: *scale* (the size and constitution of production facilities) and *intensity* (full- versus part-time production or the volume of product manufactured). Influenced by the monolithic model of craft production, they have equated the recovery of large and anomalous deposits of craft debris with nonresidential "workshops," when what they actually defined are "craftwork dumps" or unusual concentrations of production and manufacturing discards (Clark, 1989; Moholy-Nagy, 1990). Although we concur with these authors that the volume or intensity of stone tool manufacture appears unusually high, at densities clearly signaling exchange far beyond the needs of single production or domestic units, the scale most clearly associated with these production activities is not a nonresidential factory, but the household.

Alternatively, the critics of the lithic specialization interpretation at these sites (Clark, Mallory) also have been somewhat misled by the expectations of the monolithic model. At both Teotihuacan (Spence, 1981:771) and Colha (Roemer, 1982:77), lithic debitage was found mixed with quantities of domestic trash. The critics implied that this association indicates nonspecialized manufacture for immediate domestic consumption. They asserted that if the production was taking place in houses, the volume of production was by implication rather low. But the suggested domestic association of the deposits speaks *only* to the dimension of scale, not production intensity. At both Teotihuacan and Colha, rather massive quantities of production waste are suggestive of high-intensity manufacture, but there is no clear indication that this craftwork was enacted outside domestic settings. What we see then is specialized, household-scale production that yielded high volumes of craft goods at least partly for exchange.

We propose a more multidimensional perspective on Teotihuacan and Colha craft specialization. Our view decouples the attributes of scale and intensity and challenges the expectations of the monolithic model of craft specialization. At both sites, the proposed interpretation seems to fit the extant data better than either prior alternative. Yet at neither Teotihuacan nor Colha can we unequivocally link high-intensity, high-volume craftwork with domestic-scale contexts. In the remainder of this chapter, we address this problem through the firmer documentation of high-intensity domestic-scale manufacture. Our attention turns to ceramic, shell ornament, and lapidary production at the ancient Ejutla site to illustrate that specialized production (at high volumes) was enacted at the household scale in ancient Mesoamerica.

ANCIENT EJUTLA: RESEARCH BACKGROUND

Before returning to the topic of craft specialization, we introduce the Ejutla site, which is situated in the Southern Highlands of Oaxaca, Mexico, under the modern town of Ejutla de Crespo (Figure 3). In 1984 and 1985, we implemented a regional settlement pattern survey of the Ejutla Valley (Feinman and Nicholas, 1990). During this systematic survey, we mapped and recorded the prehispanic site of Ejutla, which was one of the largest prehispanic settlements in the region between the Terminal Formative and Late Postclassic periods (200 B.C.–A.D. 1520). At the eastern edge of both the modern town and the prehispanic site, we encountered unusual densities of cut and broken marine shell in 2 ha of plowed fields.

Subsequently, over four summers (1990–1993), we directed a block excavation that exposed 190 m^2 in this eastern sector of the Ejutla site, including a Classic-period (A.D. 200–800) residential structure, several ash-filled pits, dense midden areas with high concentrations of cut shell and chipped stone debris, and unusual densities of broken pottery, including numerous figurine fragments (Figure 4) (Feinman and Nicholas, 1993, 1995; Feinman et al., 1993).

The excavated structure, roughly 6 m × 4 m with a small attached work area to the north, was defined by a stone foundation and a mixed earth and crushed-bedrock floor. Several factors point to the domestic nature of this complex, including the discovery of a small subfloor tomb that included at least four individuals, the recovery of a shallow firepit and probable cooking area immediately

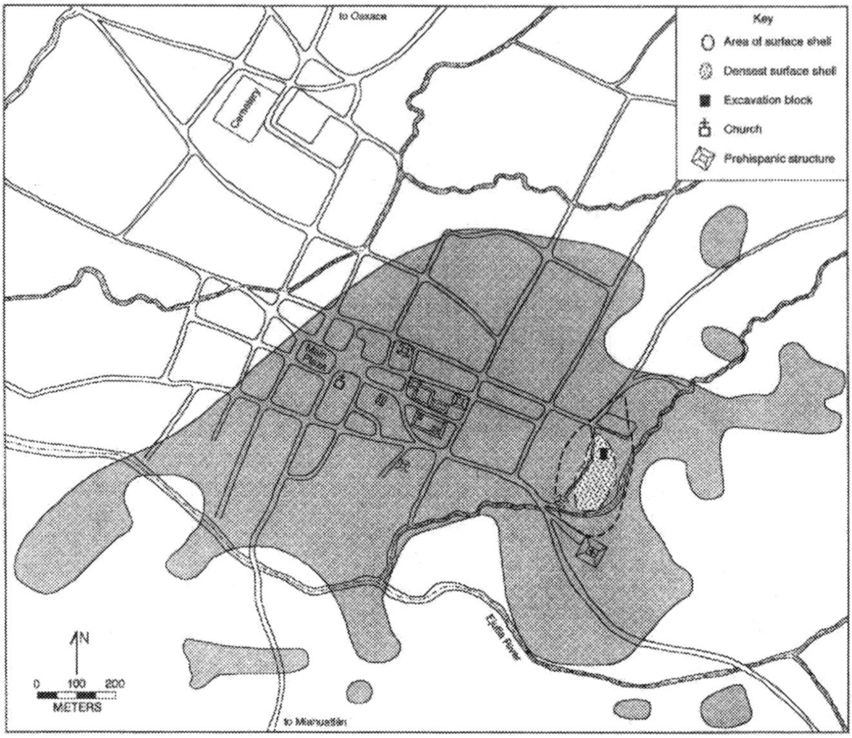

Figure 3. Ejutla de Crespo and the limits of the Ejutla site in the Classic period.

outside the structure, and the presence of food remains and other residential trash in the middens adjacent to the house. In the Classic period (e.g., Winter, 1974, 1995), residential complexes in Oaxaca were often associated with domestic tombs (on the order of the one found in Ejutla).

The modest contents of the tomb suggest that this household was not of especially high status. Yet the presence of the tomb and the size and nature of the house construction (employing a stone foundation composed of roughly finished stones rather than rough cobbles) indicate that the residents of this household were not entirely disadvantaged. Some relatively minor indications of dietary stress were evidenced on the skeletal material from the tomb. The presence of at least four individuals (three adults) in the tomb, along with the disarticulated condition of the earliest of the tomb occupants (Middleton et al., 1998), indicates that the tomb (and likely the house above it) was in use for at least a generation, if not somewhat longer.

Roughly 30 m south of the excavation block, we noted large quantities of

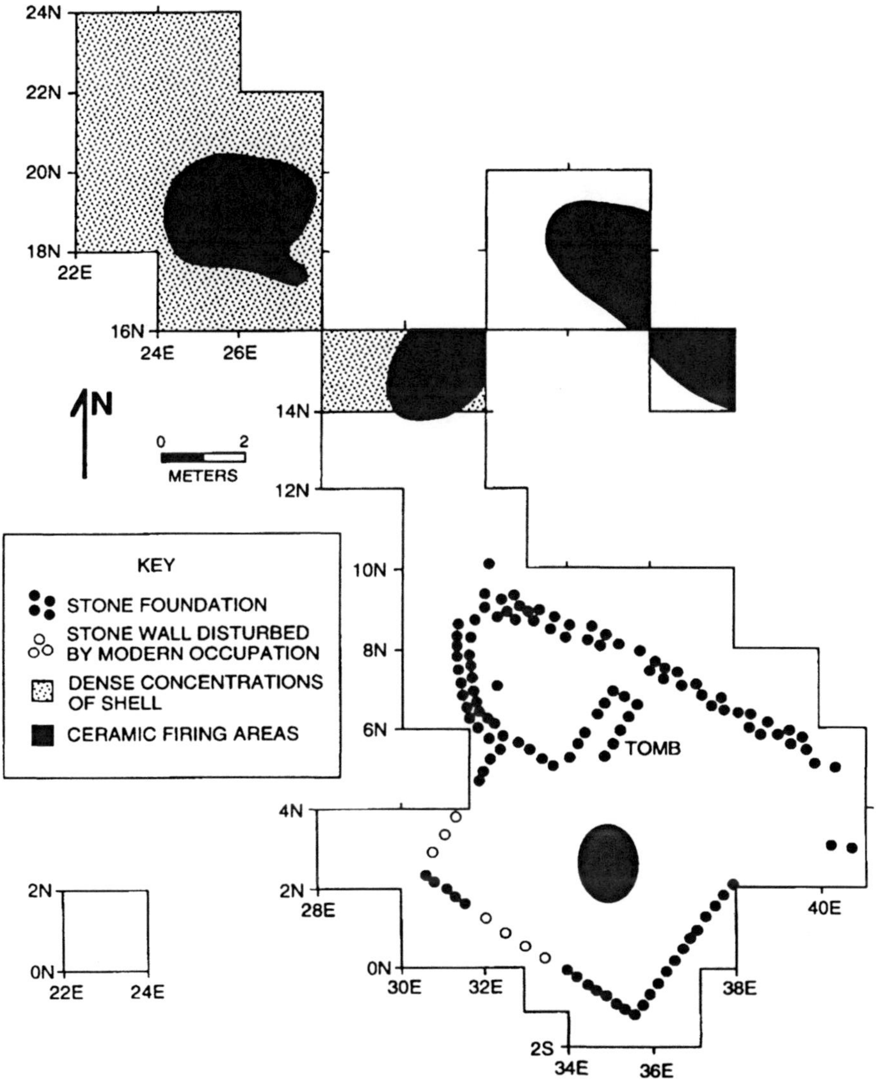

Figure 4. Plan of the Ejutla site excavations.

shell debris, worn obsidian blades, and several cut-stone foundation blocks on the surface that appear to be remains from a second structure that has been destroyed by contemporary plowing. We suspect that these two structures were part of a barrio in which the residents of the component households worked marine shell into ornaments.

SHELL ORNAMENT MANUFACTURE

More than 24,000 pieces of marine shell were recovered in the Ejutla excavations. The majority of this assemblage (about 60%) consists of broken shell and other small debris, while more than one third (35%) has clear evidence of modification or working, such as cut edges, string-cut marks, and abraded or drilled surfaces (Figure 5). The remaining 5% includes finished and unfinished ornments and unmodified whole shells. The whole shells that we recovered were from small marine varieties; no complete shells from large marine species were recovered. Yet the wide range of shell body parts represented in the debris, including intact gas-

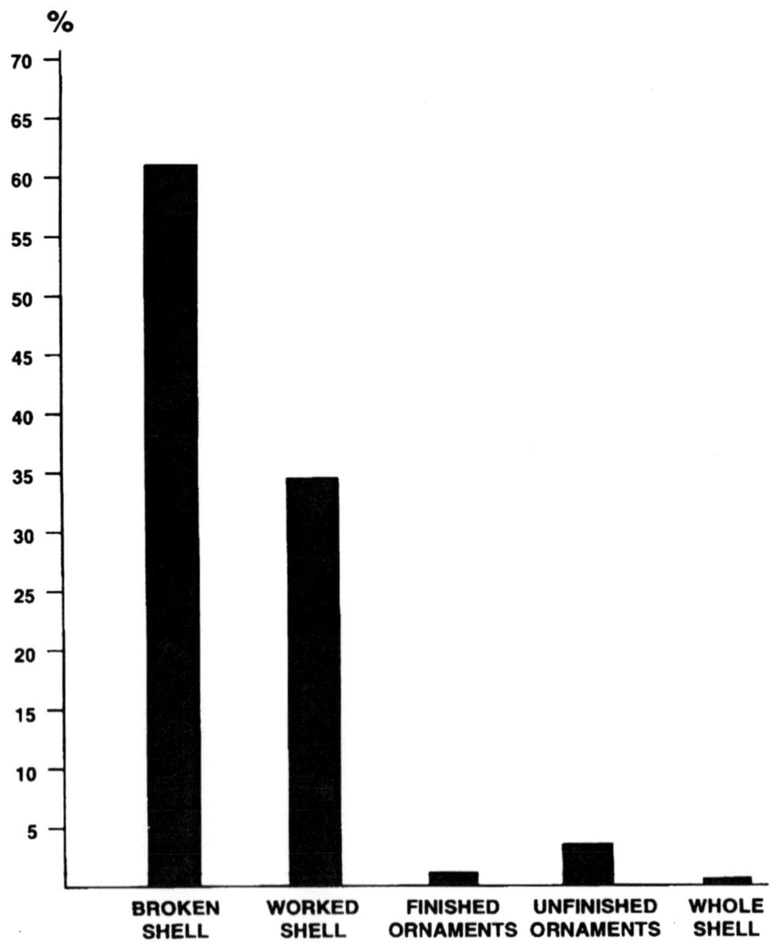

Figure 5. The shell assemblage at Ejutla.

tropod spires and columellas, large rim and body fragments, and large bivalve hinges and margins, indicates that most of the shell was brought into Ejutla as whole (or nearly complete) shells and that primary breakage and working of the shells occurred on site.

Most of the shell varieties recovered in the Ejutla excavations were native to the Pacific Coast, roughly 100 km away across high mountains. Only two identified shells were definitely of Atlantic origin. Although a diversity of shell species was represented (over 90 different taxa of marine shell have been identified), seven genera account for over 95% of the identifiable shell (Figure 6). Most of these genera (especially *Strombus, Pinctada, Spondylus*) were commonly employed to fashion ornaments across prehispanic Mesoamerica (Suárez, 1981).

The Ejutla artisans crafted a wide variety of ornaments (Figure 7), and there are relatively clear patterns of the kinds of products that were made using each shell taxon. *Pinctada mazatlanica*, the most abundant species in the Ejutla collections, was used to make the most commonly encountered ornaments—small plaques for mosaic inlay (Figure 8) and small disks (Figure 9). Although a few of the disks were perforated for stringing, most of them do not appear to have been completely finished; they likely represent an intermediate stage in ornament production.

We also recovered a variety of beads and pendants, including formed beads, where the original shape of the shell has been entirely modified through craft work, and small whole gastropods that were perforated for stringing. At Ejutla, small beads were generally made from *Spondylus* or *Chama*; *Strombus* and other

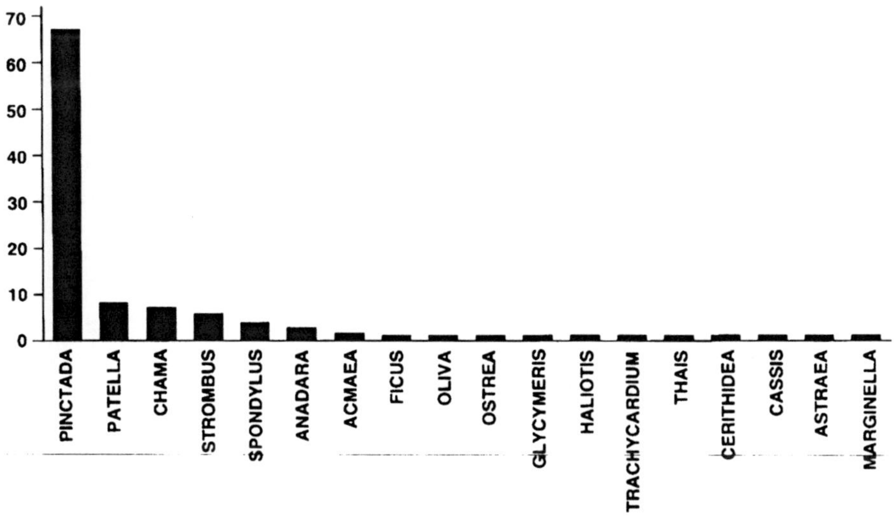

Figure 6. The most common shell genera at Ejutla.

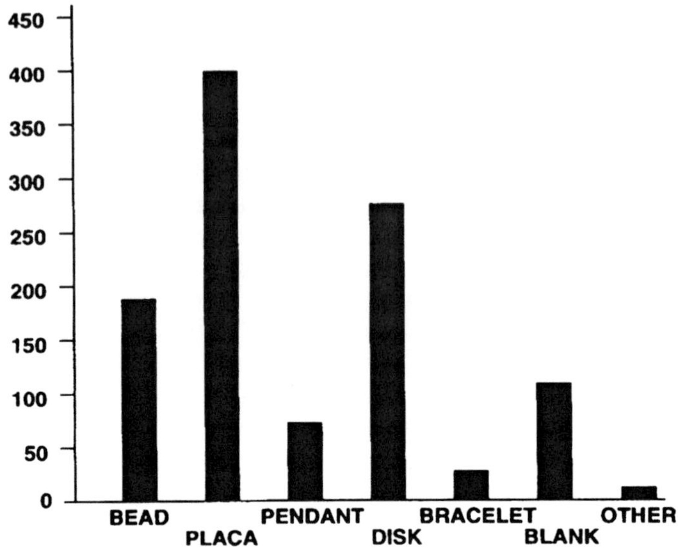

Figure 7. The range of shell ornaments at Ejutla.

larger gastropods were used to craft large beads. Most of the formed pendants were made from *Spondylus* shells. Giant limpets (*Patella mexicana*) were used mostly for making bracelets.

Through a systematic examination of the shell and stone debris, we have identified many of the shell-working techniques employed by the Ejutla craftworkers. Various stone materials were used to cut, abrade, and drill the shells. Heavily worn obsidian blades were frequently recovered in association with shell debris. These spent blades would have been effectively dulled by repeatedly cutting and working the hard, abrasive shell (see Lewenstein, 1987:113; Parry, 1987:108). Chert artifacts also were abundant, most notably as small, solid microdrills that have been linked to the perforation of beads and pendants elsewhere in Oaxaca (Parry, 1987) and in other global regions (e.g., Mester, 1985:107; Yerkes, 1989:115).

The nature of the edge on many shell ornaments and much of the debris indicates that several kinds of perishable materials also were used to process the shell. String was used in conjunction with water and an abrasive such as sand to cut small tabular pieces from the walls of various shells. This method was preferred to shape small plaques of *Pinctada mazatlanica* for use in mosaic inlay. String also was used to cut tabular ornaments from the walls of large gastropods, such as *Strombus*, and also blanks from the thicker parts of the shell, such as the

Figure 8. Plaques of shell cut from *Pinctada mazatlanica.*

base of the columella, which could be further processed into beads or other finished ornaments.

Perhaps the most distinctive technique was the use of hollow tubular drills (Foshag, 1957; Holmes, 1919), most likely made of cane (Caso, 1965), to extract small circular pieces from the walls of *Pinctada mazatlanica* (this technique was used infrequently on other kinds of shell). This technique produces standardized ornaments of fairly uniform size; the average diameter is 11 mm. These pieces could have been either used in mosaic inlay or perforated to make flat, disk beads.

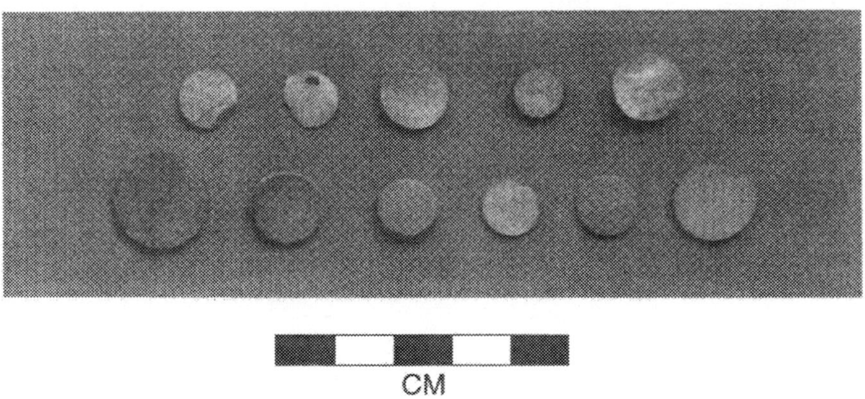

Figure 9. Disks of shell cut from *Pinctada mazatlanica.*

Subsequent analyses of marine shell artifacts from the Classic-period Valley of Oaxaca urban center of Monte Albán[1] have revealed that many of the shell ornament forms and taxa from that site mirror the forms and varieties noted in the Ejutla collections. In some instances, the shell ornaments at Monte Albán appear to be more finished versions of similar items recovered in Ejutla. Yet, although the Ejutla material includes mostly broken, incomplete, or waste pieces, the Monte Albán collections include a far higher proportion of completed artifacts (Figure 10). Finished shell ornaments constitute almost 20% of the analyzed shell from Monte Albán, compared with less than 1% of the Ejutla assemblage.

No known site in the Central Valleys of Oaxaca equals the Ejutla site in terms of the concentration of marine shell artifacts and, especially, debris. In fact, areas of shell craftwork have been identified at only two other valley sites, the centers of Monte Albán and San José Mogote. Nevertheless, shell ornaments are fairly common in elite contexts at Monte Albán and elsewhere in Oaxaca (e.g., Kolb, 1987).

In the Ejutla excavations, most of the marine shell was found in exterior midden contexts within 10 m of the residential complex. Nevertheless, two of the more finished shell ornaments recorded during the project were recovered on the floor of the house (Figure 11). Some shell debris also was found in association with the structure floor, although in much lower quantities than noted in midden areas. Microartifactual flecks of shell that were derived from the heavy fraction of soil samples taken from the floor of the house also serve to tie the working of marine shell to the domestic structure (Feinman et al., 1993).

Overall, only a small proportion of the entire Ejutla shell assemblage included finished or partially finished ornaments. Only one tiny shell bead was found in the domestic tomb. Thus it appears that the inhabitants of this part of the Ejutla site, and the occupants of the excavated house in particular, made shell ornaments for exchange. Most likely, these objects were exported beyond the limits of the settlement. In contrast, there is no empirical basis to suggest that this volume of shell ornament production (found in conjunction with so few finished pieces) was entirely intended for immediate household consumption or even exclusively local use.

CERAMIC AND FIGURINE PRODUCTION

Unexpectedly, the Ejutla excavations also yielded ample evidence for ceramic production. The initial indicator was an unusual abundance of figurines found in

[1]The Monte Albán shell collections included in this analysis were recovered during recent excavations at the site by Ernesto González Licón and Marcus Winter, both of the Instituto Nacional de Antropología e Historia in Mexico. We thank them for allowing us to examine their collections. The graphs and tabulations here reflect a synthesis of the materials from both projects.

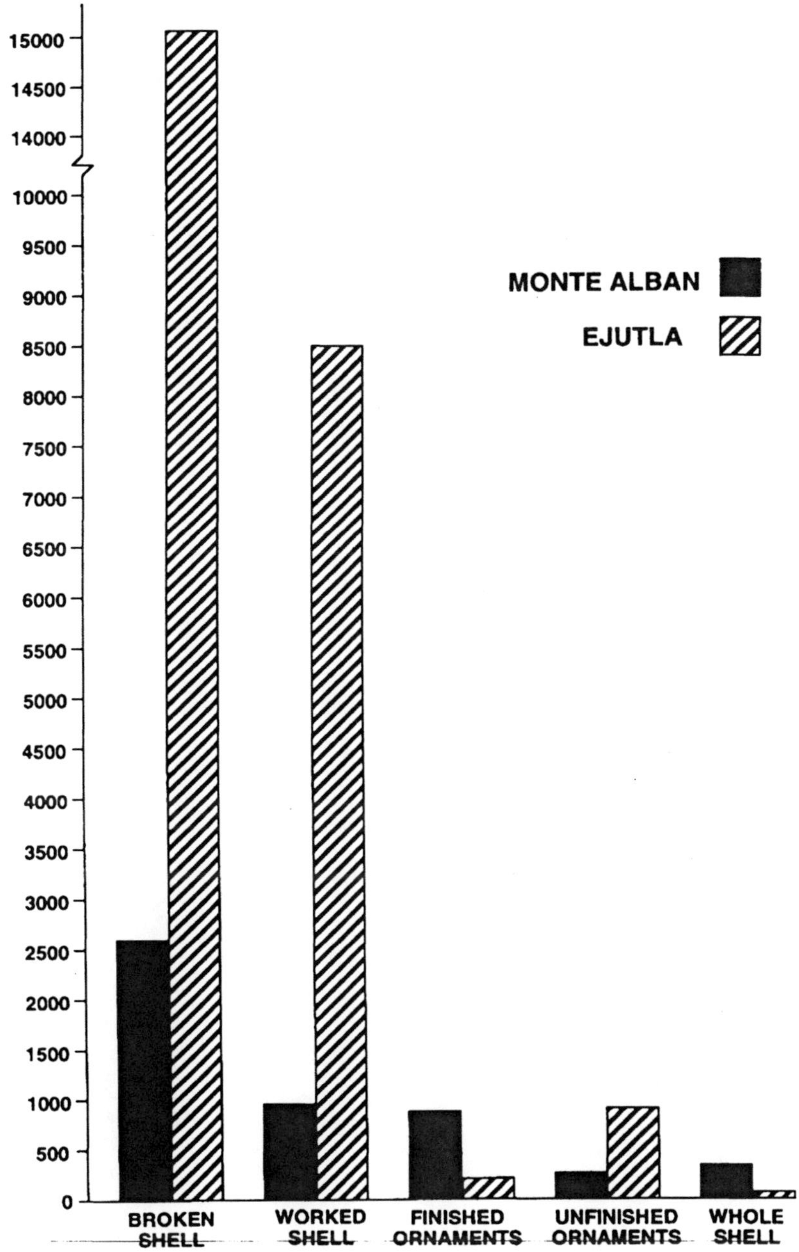

Figure 10. Comparison of the shell assemblages at Monte Albán and Ejutla.

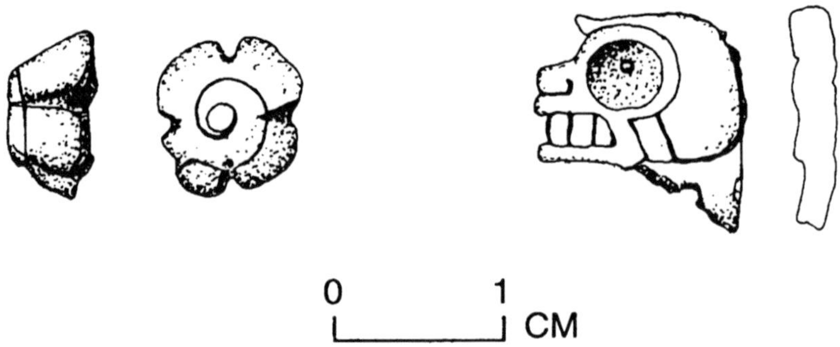

Figure 11. Two shell ornaments recovered from the floor of the excavated house in Ejutla.

midden contexts. In total, 2,000 figurines were recovered, many of them broken or malformed. More than 60 ceramic molds (at least 15 for figurines) also were recorded. Not surprisingly, some of the molds matched commonly noted figurine varieties (Figure 12). Overall, more than 900 pottery wasters were found. Almost all the categories of ceramic wasters defined by Redmond (1979) were present in the Ejutla collections.

We also excavated six shallow subsurface features that the prehispanic

Figure 12. Ceramic mold and similar figurine head from Ejutla.

Ejutleños dug into the soft bedrock. One of these features lies under (and so clearly predates) the excavated residential structure, but a number of others (just north of the structure) seem contemporary with it. These oblong features, which often were burnt at the base of the pit, are interpreted to have been used as "pit kilns" (Rye, 1981) for ceramic firing (Abascal, 1976; Bordaz, 1964; Heacock, 1995; Sheehy, 1992) (Figure 13). These features were associated with fire-cracked rock, charcoal, lenses of ash, and other indications of burning, as well as wasters, sherds that were overfired and/or fired more than once, slaglike material, and heavy densities of clay concretions (see Balkansky et al., 1997, for greater detail about these features).

At the base of one of these pits, we encountered large broken sherds, including upside-down jars, along with cobbles, which all may have been used as spacers (to separate vessels) during firing. This kiln furniture was interspersed at the base of the pit with a thick layer of ash and charcoal, reflecting an incomplete burn (Russell, 1994). In most cases, after its final use for firing, the pit kilns served as refuse depositories. The lower levels of these dumps, and the deposits surrounding the firing features, often included a number of complete, unused vessels that were flawed or broken immediately before deposition. These nearly pristine discards were similar in paste and form to the most prevalent classes of wasters (and highly abundant forms) in the Ejutla sample. Although the Ejutla potters made a wide range of forms, figurines, tortilla griddles, and incense burners were a few of their mainstays. The emphasis on these forms is significant because a household does not require huge numbers of incense burners or tortilla griddles.

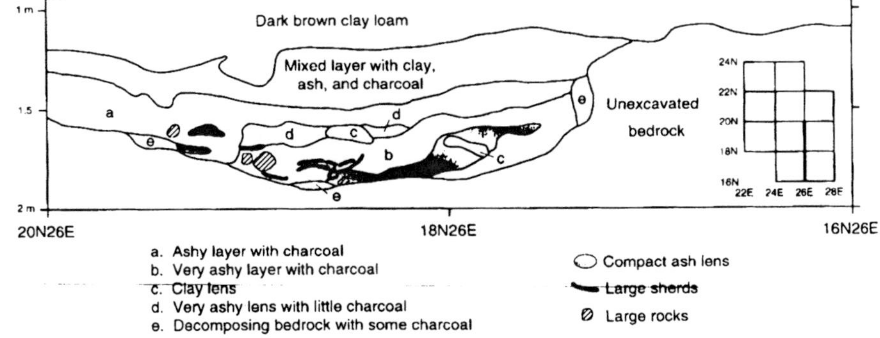

Figure 13. Profile of Ejutla pit kiln.

The abundance of ceramic wasters in our excavations, in conjunction with the presence of figurine molds, clay concretions, and the excavated pit kilns, securely point to pottery production. Determination of the intensity of prehispanic pottery manufacture is more difficult. However, in our modestly sized excavation block, we recovered an astounding quantity of ceramic artifacts. We analyzed more than 210,000 pottery fragments, which had a combined weight exceeding 3500 kg.

The form and paste of ceramic artifacts above, below, and associated directly with the floor of the structure enable us to link the Ejutla house with the immediately adjacent midden and firing areas. Preliminary petrographic and elemental (ICP) analyses of raw clays taken from the current site surface itself (and from a finer clay bed in a nearby barranca) are qualitatively (mineralogically and elementally) similar to the pastes of Ejutla figurines and other vessels (Carpenter and Feinman, 1999). Compositionally, a fired clay concretion found during excavation was also similar to the raw clays. Interestingly, in contrast to the figurines, for which we find lots of large inclusions in the paste, some locally produced Ejutla pottery bowls appear to have been made with processed clays that were significantly finer than the available raw clays. Andrea Carpenter has been able to reproduce these finer pastes by beating and then levigating the raw Ejutla clays.

Spatial proximity between the domestic complex, the firing features, and the midden areas (loaded with ceramic manufacturing debris), as well as artifactual associations across these features, links the Ejutla household with ceramic manufacturing. Yet, could this domestic production have been entirely for household use? Although the evidence against this proposition is not as strong as it is for the relatively rare and exotic good, shell, the rarity of finished products, the abundance of wasters, the specific ceramic forms that were manufactured, and comparative analyses with figurines found at neighboring sites all point to production for exchange.

THE SCALE AND INTENSITY OF EJUTLA CRAFTWORK

To this point, we have described domestic-scale production in which shell ornaments and a range of ceramic forms were likely made, at least in part, to trade. In the excavated middens, craft debris often was found with domestic trash, such as food remains and sherds from charred cooking jars. Such remains further tie the craft debris to a domestic context. The location of the craft middens adjacent to the excavated structure also conforms with ethnoarchaeological findings that garbage created by residential units generally is deposited on the house lot, but in areas that do not interrupt other household activities (Hayden and Cannon, 1983).

At the same time, finished products in both materials (shell and ceramic) seem to have been traded beyond the household of the producers. Given the relative uniqueness of the Ejutla shell concentration for highland Oaxaca, shell artifacts were likely exchanged more widely than the ceramic goods. Both crafts entailed the knowledge and application of specialized technologies, including the firepits and molds for pottery, and hollow tubular (cane) drills and chert microdrills for the shell (Feinman and Nicholas, 1995). Analysis of the exterior midden deposits has revealed that the chert microdrills were apparently finished close to the time that shell ornament manufacture was carried out, because the reduction debris from both tasks was almost always depositionally mixed. The chert for the microdrills was quarried from a source roughly 3 km from the Ejutla house; however, we do not know whether it was procured directly or through exchange with the occupants of a small contemporary settlement situated closer to the stone source.

Microartifactual analysis (Feinman et al., 1993; Middleton, 1998) also has revealed tiny chert flakes as well as microscopic indicators of lapidary work on the structure floor. Rectangular onyx plaques and cylindrical drill cores (Figure 14), greenstone chunks and flakes, and flakes and chunks of nonlocal stones also were recovered in the excavations, principally in the exterior midden. The volume of these stone materials was much lower than the quantities of shell and pottery

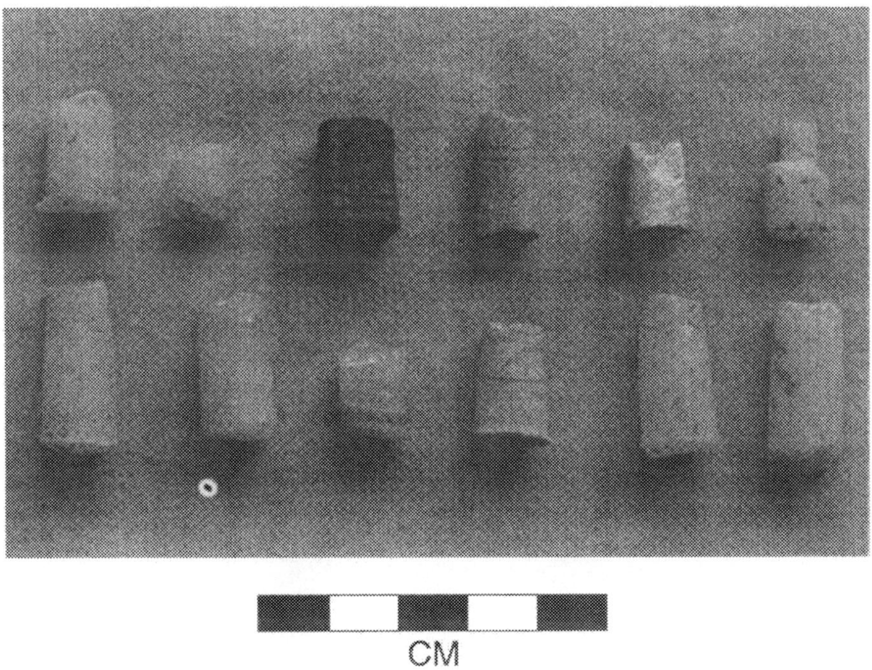

Figure 14. Onyx drill plugs from Ejutla.

debris, and no fully finished stone ornaments were recovered. Consequently, lapidary manufacture appears to have been of lesser importance and lower intensity than the other crafts. Nevertheless, the same hollow tubular drill technology was employed in processing both shell and onyx. The diameter of the majority of the onyx drill plugs matched that of many flat shell disks. Use of these "intersecting technologies" (Earle, 1994:455; Hagstrum, 1992) further supports the inference that this Ejutla household was involved in several craft activities or "multicrafting." This finding contradicts the oft-held notion in Mesoamerican archaeology that each household or community of specialists focused on a single craft.

A fundamental question then is how much involvement did the people in this household have in noncraft activities, primarily agriculture? Because we lack any agricultural features adjacent to the excavated house, this query is next to impossible to address directly. The Ejutla excavations did yield 18 polished axe fragments, including one small piece associated with the house floor. Yet polished axes could have many functions, only some of which are related to farming. For example, pottery manufacture certainly necessitated the procurement of wood for firing, which likely required an axe. No other artifacts clearly indicative of agricultural work were found. In sum, there is nothing definitive to rule out agriculture as part of this household's annual time budget, but likewise there is no reason to presume that it was the major component in its yearly labor allocation. It seems probable that household members participated in various crafts as well as some agricultural activities. Yet clearly, much labor was devoted to craft-related work.

We do not know which members of the Ejutla households practiced which specific crafts. At present, it remains possible that males and females carried out different crafts. Yet the intersecting technologies seem to argue against a strict material division. Nevertheless, the key point remains that several crafts seem to have been made for exchange in this domestic setting.

CRAFT PRODUCTION: DISCUSSION AND IMPLICATIONS

As at Teotihuacan and Colha, the Ejutla findings document relatively high-intensity craft manufacture (for exchange) that was enacted at the domestic scale. In contrast to the expectations of the monolithic model, household-scale craft production was not necessarily a secondary activity supplemental to the primary economic pursuit of agriculture. The residues of craftwork were diverse, more abundant than any indicators of agricultural work. Likewise, in opposition to the tenets of the monolithic model, the Ejutla craftworkers did not produce exclusively for themselves or their close neighbors. Finished shell objects may have been traded to places as distant as Monte Albán. They also did not rely solely on local resources. In addition, the ancient Ejutleños created huge quantities of debris, committed a significant time investment, and worked with various specialized and labor-intensive technologies. In sum, the nature of household craft production in Ejutla bears little similarity to the expectations for such manufacture outlined by the monolithic model.

Although the Ejutla findings do not conform to the monolithic model's expectations about household craft specialization, they do correspond with previous empirical findings about the general character of craft manufacture in prehispanic Mesoamerica. As at Teotihuacan and Colha, most Mesoamerican craftwork can be linked to production at the household level. Mesoamerican archaeologists have frequently described artifactual indications of high-intensity and finely skilled craft specialization, and often these findings are directly associated with residential structures or other domestic remains (e.g., Charlton et al., 1993; Healan, 1986).

Despite several tentative propositions, true workshops or factories have yet to be conclusively reported for prehispanic Mesoamerica. The familiar claims for nondomestic production in Mesoamerica are based principally on indicators of high intensity (production in volume) rather than independent measures of scale itself (e.g., Santley et al., 1989). For prehispanic Mesoamerica, we lack much definitive evidence for workshop structures or fabrication areas that were not directly associated with domestic quarters.

The picture from ethnohistoric documents is similar, lacking clear evidence for nondomestic workshops of significant scale. The Florentine Codex (Sahagún, 1950-1963) illustrates craftwork in detail, but it appears to have been carried out either in domestic or "attached" palace settings with only a small number of participating specialists. Even in rural Mesoamerica today, full-time specialists, producing large volumes of craft goods for outside consumption and markets, often work in domestic settings, not factories (e.g., Hendry, 1992). Consequently, for prehispanic Mesoamerica, the broadly assumed isomorphism between scale and intensity seems to find little support. To understand ancient Mesoamerican craft specialization and the nature of the region's prehispanic economy, the scale and intensity of craft specialization should be examined as discrete dimensions.

This discussion is not meant to imply that nondomestic workshops will never be defined securely for ancient Mesoamerica. In fact, we fully expect that such workshops will be found in a few special settings, but to identify them conclusively will require the opening of large excavation blocks. And we expect that even when we have exposed more such excavation blocks, nonresidential workshops will be rare in prehispanic Mesoamerica. Alternatively, what this also implies is that when we find high volumes of craft manufacturing debris in domestic contexts, we cannot immediately presume that it represents only the residues of a part-time industry and/or low-intensity production. In ancient and modern Mesoamerica, intensive craftwork can and did occur in domestic settings.

IMPLICATIONS FOR ANCIENT MESOAMERICAN ECONOMIES

To this point, we have proposed that a significant amount of ancient Mesoamerican craft production occurred in domestic contexts, that this household manufacture often involved the practice of multiple crafts, and that craft items could be manufactured in such contexts at high intensities for exchange. This modification to the monolithic model has several important ramifications for previous interpretive models of prehispanic Mesoamerican economies, as well as for future archaeological fieldwork agendas.

To date, energy-driven models of ancient Mesoamerican craft manufacture have tended to minimize production for exchange in most prehispanic contexts (e.g., Sanders and Webster, 1988). These analyses have postulated that the limited production and transportation technologies available in Mesoamerica would have constrained the capabilities of most craftworkers (the exception being obsidian tool production) to make a living, because these specialized producers could not supply an adequate number of consumers. Yet such energetic models tend to presume that artisans specialized only in a single craft. The specialized production of a range of goods by single-craft households would remove some of the presumed transportation constraints on producers, because consuming households could procure a range of different craft goods from the same specialist household. Domestic production for exchange would thereby become a more feasible economic strategy.

Likewise, if high-intensity production for exchange was ongoing in Mesoamerican households, then the notion that most Mesoamerican cities were "regal-ritual" centers without significant economic functions also would be called into question (Sanders and Webster, 1988). In the regal-ritual model, craft production in most ancient Mesoamerican economies (the regional settlement systems dominated by Teotihuacan, Tula, and Tenochtitlán are exceptions) was restricted to elite goods produced by attached specialists or was carried out at low intensities in domestic contexts by rural producers. Yet this does not seem to conform with the published evidence from large sites like Tikal (Becker, 1973), Xochicalco (Hirth, 1995), Otumba (Charlton et al., 1993), and Monte Albán (Blanton, 1978), where anomalous concentrations of craft debris (ceramic production, stone tool working) have been noted, albeit generally in domestic contexts. Because we can no longer presume that household production was by definition low in intensity and economically insignificant, and the volumes of recovered debris were considerable in each of these cases, it seems reasonable to acknowledge that the economic importance of nonagricultural production in most ancient Mesoamerican cities was greater than the regal-ritual model leads us to suspect. Utilitarian craft production was likely often a key feature of prehispanic Mesoamerican cities, but much of this specialized craft activity may have taken place in houses (rather than in nonresidential workshops).

More than three decades ago in his seminal study of early Mesoamerican villages, Kent Flannery (Flannery and Winter, 1976) wisely directed his colleagues to study household activities and the ways they were similar and different across communities, regions, and even the entire Mesoamerican macroregion. For his primary focus of investigation, the Formative period (1500 B.C.–A.D. 200), Flannery established that there were some activities present in all houses, whereas other activities were more differentially and exclusively distributed across sites and larger geographic entities. By documenting the differences and similarities between household activities, Flannery developed a foundation for understanding how Formative-period Mesoamerican households were interconnected into larger economic and political units.

Based on the findings from our study, it now appears that Flannery's domestic vantage also may be crucial for understanding later (Classic and Postclassic) prehispanic societies. If most ancient Mesoamerican economic production occurred in domestic contexts, then the basis for explaining change and diversity in prehispanic Mesoamerican societies lies in understanding the myriad ways in which prehispanic households deployed their labor and then were articulated into larger economic and political networks. In other words, Flannery's archaeological "roadmap" for the Formative period provides a productive guide for what we need to know about later periods in ancient Mesoamerica as well.

Unfortunately, despite the passage of time, large samples of excavated houses from later prehispanic Mesoamerican sites remain relatively scarce. We can only begin to address the kinds of issues for the Classic and Postclassic periods that Flannery considered three decades ago for the Formative. In Mesoamerica, there is an urgent need to implement the field research designs pioneered in *The Early Mesoamerican Village* (Flannery, 1976), such as large lateral uncoverages and the careful excavation of meaningful behavioral entities, like houses. In other words, if our goal is to understand variation and change in ancient Mesoamerican economies, then we desperately require more house excavations, even if they are undertaken in lieu of the more frequent uncoverage of isolated test pits, tombs, and temples. If 30 years from now, the archaeological findings from these kinds of domestic studies undertaken at cities, towns, and villages across Mesoamerica are published, then the discipline (and we its practioners) will have left Kent V. Flannery the scientific legacy that he so richly merits.

ACKNOWLEDGMENTS

The National Science Foundation (BNS 89-19164, BNS 91-05780, and SBR 9304258 to the senior author) has provided the principal support for the Ejutla excavations. Additional funding has been generously awarded by the H. J. Heinz Foundation, the National Geographic Society, the Graduate School of the University of Wisconsin, Madison, and Arvin B. Weinstein. The Instituto Nacional de

Antropología e Historia has provided essential permissions and support. We also wish to thank the members of our 1990–1993 field and laboratory crews for their diligent efforts and endurance. We also thank Paula Schulze and Jennifer Veser for drawing the shell artifacts in Figure 11 and Lane Fargher for assisting in the preparation of Figures 3 and 13.

REFERENCES

Abascal, R., 1976, Los primeros pueblos alfareros prehispánicos. *Comunicaciones, Suplemento, Proyecto Puebla-Tlaxcala*, III:49–52. Fundación Alemana para la Investigación Científica, Puebla.

Balkansky, A. K., Feinman, G. M., and Nicholas, L. M., 1997, Pottery Kilns of Ancient Ejutla, Oaxaca, Mexico. *Journal of Field Archaeology* 24:139–160.

Becker, M. J., 1973, Archaeological Evidence for Occupational Specialization among the Classic Maya at Tikal, Guatemala. *American Antiquity* 38:396–406.

Blanton, R. E., 1978, *Monte Albán: Settlement Patterns at the Ancient Zapotec Capital*. Academic Press, New York.

Bordaz, J., 1964, Pre-Columbian Ceramic Kilns at Peñitas, a Post-Classic Site in Coastal Nayarit, Mexico. Ph.D. dissertation, Department of Anthropology, Columbia University, New York.

Carpenter, A., and Feinman, G. M., 1999, The Effects of Behaviour on Ceramic Composition: Implications for the Definition of Production Locations. *Journal of Archaeological Science* 26:783–795.

Caso, A., 1965, Lapidary Work, Goldwork, and Copperwork from Oaxaca. In *Handbook of Middle American Indians*, Volume 3, Part 2, edited by G. R. Willey, pp. 896–930. University of Texas Press, Austin.

Charlton, C. O., Charlton, T. H., and Nichols, D. L., 1993, Aztec Household-Based Craft Production: Archaeological Evidence from the City-State of Otumba, Mexico. In *Prehispanic Domestic Units in Western Mesoamerica: Studies of Household, Compound, and Residence*, edited by R. S. Santley and K. G. Hirth, pp. 147–171. CRC Press, Boca Raton.

Clark, J. E., 1989, Hacia una definición de talleres. In *La obsidiana en Mesoamérica*, edited by M. Gaxiola G. and J. E. Clark, pp. 213–217. Instituto Nacional de Antropología e Historia, Mexico.

Costin, C. L., 1991, Craft Specialization: Issues in Defining, Documenting, and Explaining the Organization of Production. *Archaeological Method and Theory* 3:1–56.

Earle, T. K., 1994, Wealth Finance in the Inka Empire: Evidence from the Calchaqui Valley, Argentina. *American Antiquity* 59:443–460.

Feinman, G. M., and Nicholas, L. M., 1990, At the Margins of the Monte Albán State: Settlement Patterns in the Ejutla Valley, Oaxaca, Mexico. *Latin American Antiquity* 1:216–246.

Feinman, G. M., and Nicholas, L. M., 1993, Shell-Ornament Production in Ejutla: Implications for Highland-Coastal Interaction in Ancient Oaxaca. *Ancient Mesoamerica* 4:103–119.

Feinman, G. M., and Nicholas, L. M., 1995, Household Craft Specialization and Shell Ornament Manufacture in Ejutla, Mexico. *Expedition* 37(2):14–24.

Feinman, G. M., Nicholas, L. M., and Middleton, W. M., 1993, Craft Activities at the Prehispanic Ejutla Site, Oaxaca, Mexico. *Mexicon* 15:33–41.

Flannery, K. V., and Winter, M. C., 1976, Analyzing Household Activities. In *The Early Mesoamerican Village*, edited by K. V. Flannery, pp. 34–44. Academic Press, New York.

Flannery, K. V. (editor), 1976, *The Early Mesoamerican Village*. Academic Press, New York.

Foshag, W. F., 1957, *Mineralogical Studies on Guatemalan Jade*. Smithsonian Miscellaneous Collections Vol. 135(2). Smithsonian Institution, Washington, D.C.

Hagstrum, M., 1992, Intersecting Technologies: Ceramic Tools for Inka Metallurgy. Paper presented at the 28th Annual Symposium on Archaeometry, Los Angeles.

Hayden, B., and Cannon, A., 1983, Where the Garbage Goes: Refuse Disposal in the Maya Highlands. *Journal of Anthropological Archaeology* 2:117–163.

Heacock, L. A., 1995, Archaeological Investigations of Three Mesa Verde Anasazi Pit Kilns. *Kiva* 60:391–410.

Healan, D. M., 1986, Technological and Nontechnological Aspects of an Obsidian Workshop Excavated at Tula, Hildago. In *Research in Economic Anthropology, Supplement 2: Economic Aspects of Prehispanic Highland Mexico*, edited by B. L. Isaac, pp. 133–154. JAI Press, Greenwich, CT.

Hendry, J. C., 1992, *Atzompa: A Pottery Producing Village of Southern Mexico in the Mid-1950's*. Publications in Anthropology No. 40. Vanderbilt University, Nashville.

Hester, T. R., and Shafer, H. J., 1994, The Ancient Maya Craft Community at Colha, Belize, and Its External Relationships. In *Archaeological Views of the Countryside: Village Communities in Early Complex Societies*, edited by G. M. Schwartz and S. E. Falconer, pp. 48–63. Smithsonian Institution Press, Washington, D.C.

Hirth, K. G., 1995, Urbanism, Militarism, and Architectural Design: An Analysis of Epiclassic Sociopolitical Structures at Xochicalco. *Ancient Mesoamerica* 6:237–250.

Holmes, W. H., 1919, Abrading Processes. In *Handbook of Aboriginal American Antiquities, Part I: The lithic Industries*, pp. 344–357. Bureau of American Ethnology, Bulletin No. 60. Smithsonian Institution, Washington, D.C.

Kolb, C. C., 1987, *Marine Shell Trade and Classic Teotihuacan, Mexico*. BAR International Series No. 364. British Archaeological Reports, Oxford.

Lewenstein, S. M., 1987, *Stone Tool Use at Cerros: The Ethnoarchaeological and Use-Wear Evidence*. University of Texas Press, Austin.

McAnany, P. A., 1993, Resources, Specialization, and Exchange in the Maya Lowlands. In *The American Southwest and Mesoamerica: Systems of Prehistoric Exchange*, edited by J. E. Ericson and T. G. Baugh, pp. 213–245. Plenum, New York.

Mester, A. M., 1985, Un taller manteno de la concha madre perla del sitio Los Frailes, Manabi. *Miscelánea Antropológica Ecuatoriana* 5:101–111.

Middleton, W. D., 1998, Craft Specialization at Ejutla, Oaxaca, Mexico: An Archaeometric Study of the Organization of Household Craft Production. Ph.D. dissertation, Department of Anthropology, University of Wisconsin, Madison.

Middleton, W. D., Feinman, G. M., and Molina Villegas, G., 1998, Tomb Use and Reuse in Oaxaca, Mexico. *Ancient Mesoamerica* 9:1–11.

Moholy-Nagy, H., 1990, The Misidentification of Mesoamerican Lithic Workshops. *Latin American Antiquity* 1:268–279.

Parry, W. J., 1987, *Chipped Stone Tools in Formative Oaxaca, Mexico: Their Procurement, Production, and Use*. Memoirs No. 20. Museum of Anthropology, University of Michigan, Ann Arbor.

Peacock, D. P. S., 1982, *Pottery in the Roman World: An Ethnoarchaeological Approach*. Longman, London.

Redmond, E. M., 1979, A Terminal Formative Ceramic Workshop in the Tehuacán Valley. In *Prehistoric Social, Political, and Economic Development in the Area of the Tehuacán Valley: Some Results of the Palo Blanco Project*, edited by R. D. Drennan, pp. 111–127. Technical Reports No. 11. Museum of Anthropology, University of Michigan, Ann Arbor.

Roemer, E., 1982, Investigation at Four Lithic Workshops at Colha, Belize: 1981 Season. In *Archaeology at Colha, Belize: The 1981 Interim Report*, edited by T. R. Hester, H. J. Shafer, and J. D. Eaton, pp. 75–84. Center for Archaeological Research, University of Texas, San Antonio.

Russell, G. S., 1994, Cerro Mayal, Peru: Moche Ceramic Workshop Excavated. *Backdirt* Spring:6–7.

Rye, O. S., 1981, *Pottery Technology: Principles and Reconstruction*. Manuals on Archaeology No. 4. Taraxacum, Washington, D.C.

Sahagún, B. de, 1950–1963, *General History of the Things of New Spain* (Florentine Codex). 12 vols. Translated by A. Anderson and C. Dibble. School of American Research, Santa Fe, and University of Utah, Salt Lake City.

Sanders, W. T., and Webster, D., 1988, The Mesoamerican Urban Tradition. *American Anthropologist* 90:521–546.

Santley, R. S., 1983, Obsidian Trade and Teotihuacan Influence in Mesoamerica. In *Interdisciplinary Approaches to the Study of Highland-Lowland Interaction*, edited by A. Miller, pp. 69–123. Dumbarton Oaks, Washington, D.C.

Santley, R. S., Arnold, P. J., and Pool, C. A., 1989, The Ceramics Production System at Matacapan, Veracruz, Mexico. *Journal of Field Archaeology* 16:107–132.

Shafer, H. J., and Hester, T. R., 1983, Ancient Maya Chert Workshops in Northern Belize, Central America. *American Antiquity* 48:519–543.

Shafer, H. J., and Hester, T. R., 1986, Maya Stone-Tool Craft Specialization and Production at Colha, Belize: Reply to Mallory. *American Antiquity* 51:158–166.

Shafer, H. J., and Hester, T. R., 1991, Lithic Craft Specialization and Product Distribution at the Maya Site of Colha, Belize. *World Archaeology* 23:79–97.

Sheehy, J. J., 1992, Ceramic Production in Ancient Teotihuacan, Mexico: A Case Study of Tlajinga 33. Ph.D. dissertation, Department of Anthropology, Pennsylvania State University, University Park.

Spence, M. W., 1967, The Obsidian Industry of Teotihuacan. *American Antiquity* 32:507–514.

Spence, M. W., 1981, Obsidian Production and the State in Teotihuacan. *American Antiquity* 46:769–788.

Spence, M. W., 1987, The Scale and Structure of Obsidian Production in Teotihuacan. In *Teotihuacan: Nuevos datos, nuevas síntesis, nuevos problemas*, edited by E. McClung de Tapia and E. C. Rattray, pp. 429–450. Universidad Nacional Autónoma de México, Mexico.

Suárez, L., 1981, *Técnicas prehispánicas en los objetos de concha*. Colección Científica 14. Instituto Nacional de Antropología e Historia, Mexico.

van der Leeuw, S., 1976, *Studies in the Technology of Ancient Pottery*. Organization for the Advancement of Pure Research, Amsterdam.

van der Leeuw, S., 1977, Towards a Study of the Economics of Pottery Making. In *Ex Horreo*, edited by B. L. Beek, R. W. Brant, and W. Gruenman van Watteringe, pp. 68–76. Cingvla 4. Albert Egges van Giffen Instituut voor Prae- en Protohistoire, University of Amsterdam, Amsterdam.

Winter, M. C., 1974, Residential Patterns at Monte Albán, Oaxaca, Mexico. *Science* 186:981–987.

Winter, M. C. (editor), 1995, *Entierros humanos de Monte Albán: dos estudios*. Centro INAH, Oaxaca.

Yerkes, R. W., 1989, Shell Bead Production and Exchange in Prehistoric Mississippian Populations. In *The Proceedings of the 1986 Shell Bead Conference: Selected Papers*, edited by C. F. Hayes, III, and L. Ceci, pp. 113–123. Research Records No. 20. Rochester Museum and Science Center, Rochester.

Part III

Political Complexity and Its Emergence

The four chapters in this part consider different aspects of sociopolitical organization and change in ancient chiefdoms and states. Collectively, the foci of these chapters dovetail with themes of chiefly and state organization (and cross-cultural variation between such polities) that have long been evident in Flannery's writings. The chapters that follow here include a multiscalar case example focused on the Formative period in the Southern Highlands of Mexico (Spencer and Redmond) that clearly has broad analytical implications, along with three largely theoretical discussions by Drennan, Wright, and Frangipane.

Drennan's work represents a thoughtful and balanced theoretical consideration of the now-popular agency models that have been proposed to account for the emergence of social inequality. His discussion points to a series of issues that need to be considered in future research on this theme. Wright also offers a reconsideration and expansion of earlier theoretical models that concern chiefdoms. In contrast to Drennan, Wright's concern is more regional in focus with an emphasis on settlement patterns and their variation. Drawing on data from several different regions, he argues that the settlement patterns of complex chiefdoms vary depending on the nature of the economy and whether tribute was paid primarily in labor or goods. Both Wright's and Drennan's chapters thematically crosscut the considerations of Frangipane, who takes a theoretical look at shifting exchange relations in the ancient Near East. Like Wright, Frangipane is concerned with the economic articulations between emergent centers and their surrounding settlements and the ways these relations vary through time and space. In concert with Drennan, Frangipane considers the ever-present administrative dialectic between "self-serving" and "system-serving" strategies and goals. In so doing, her work presents a valuable corollary to the ongoing scholarly effort to make the old catchall "redistribution" more precise and theoretically useful.

As in Part II, the authors take theoretically comparative and analytically holistic tacks to address both the processes of long-term change and, more generally, key issues in cultural evolution. Likewise, each of the chapters in this part (even those with a somewhat more defined geographic focus—Spencer and Redmond and Frangipane) endeavors to contribute to and reflect on broad cross-cultural themes. Along with the chapters in Part II, these authors illustrate some of the principal dialogues and issues that remain central to contemporary cultural evolutionary approaches as well as some of the ongoing field research and theoretical modeling being undertaken to investigate that agenda.

Chapter **6**

Lightning and Jaguars

Iconography, Ideology, and Politics in Formative Cuicatlán, Oaxaca

CHARLES S. SPENCER AND ELSA M. REDMOND

INTRODUCTION

It is not uncommon these days for archaeologists to be challenged to take a stand on whether political development is driven by material *or* ideological factors (e.g., Demarest, 1989; Demarest and Conrad, 1993; Hodder, 1991; Miller and Tilley, 1984; Shanks and Tilley, 1987). We doubt, however, that polarizing the debate in this fashion leads to a better understanding of political dynamics in the ancient past. In the present chapter we eschew such polarization and strive instead to take a comprehensive, interactional approach to *politics*, which we consider to be the strategies and tactics that people use to attain and exercise *power*, the capacity to determine the behavior of others.

Our central argument is that power and politics are multifaceted phenomena, whose manifestations usually involve a complex and often subtle blend of *both* ideological and material elements. For example, preferential *resource control* is a tactic that can help an aspiring leader impose his will on others; but followers could well decide to depose or abandon a leader whose authority is not sufficiently reinforced by ideas and beliefs that justify his material advantage. On the other hand, *ideology* can provide the justification and moral legitimacy for cen-

CHARLES S. SPENCER and ELSA M. REDMOND • Department of Anthropology, American Museum of Natural History, New York, New York 10024.

Cultural Evolution: Contemporary Viewpoints, edited by Gary M. Feinman and Linda Manzanilla. Kluwer Academic/Plenum Publishers, 2000.

tralized authority and socioeconomic inequality; but ideas can have an impact on potential followers only if they are brought out of the mental domain and put into action—through appropriate ceremony, effective symbolism, or skillful propaganda. We contend that ideology must be *actualized* (or *materialized*) in the real world to have political consequences (see DeMarrais et al., 1996). This view of ideology draws inspiration from the action-theoretic approach that Flannery and Marcus (Flannery and Marcus, 1993; Marcus and Flannery, 1996) have advocated. Of particular relevance is their definition of ideology as "the body of doctrine, myth, and symbolism of a social movement, institution, class or group of individuals, often with reference to some political or cultural plan, along with the strategies for putting the doctrine into operation" (Flannery and Marcus, 1993:263). At the same time, the ultimate legitimacy of an ideology is often achieved through *sanctification*, whereby the ideological doctrine, plan, and action all become tightly linked to—in effect, a part of—*religion*, a "specific set of beliefs in a divine or superhuman power or powers, to be obeyed and worshipped" (Flannery and Marcus, 1993:262).

TOWARD A HOLISTIC ARCHAEOLOGY

For more than 25 years, Kent Flannery has argued the case for an inclusive, multivariate approach to archaeological research. In his landmark 1972 paper "The Cultural Evolution of Civilizations," he lamented the fact that scholars investigating the evolution of complex societies had become polarized into two opposing camps, which he called the "ecologists" and the "humanists" (1972:400). Noting that modern ecologists ought to be examining the exchange of matter, energy, *and* information between populations and their environments (both natural and social), Flannery asserted:

> until now, it has mainly been the humanists who have studied the informational aspects of complex societies—art, religion, ritual, writing systems, and so on. The "ecologists" have largely contented themselves with studying exchanges of matter and energy. (1972:400)

Such polarization has led to incomplete, unsatisfying explanations of culture change, a situation that Flannery has sought to remedy by developing an expanded ecological perspective that encompassed ideas, religion, ritual, writing, art, and sociopolitical institutions along with subsistence activities, craft production, demography, and other technoeconomic factors. He argued that:

> humanists must cease thinking that ecology "dehumanizes" history, and ecologists must cease to regard art, religion, and ideology as mere "epiphenomena" without causal significance. In an ecosystem approach to the analysis of human societies, everything which transmits information is within the province of ecology. (1972:400)

Exemplifying this approach was the 1976 paper "Formative Oaxaca and the Zapotec Cosmos," in which Flannery and Marcus drew on archaeological, ethnohistorical, and ethnographic data from the Oaxaca area to analyze Zapotec ideology, ritual, social organization, and ecology over many centuries of cultural development. A central objective of the exercise, they noted, was "to find a framework for analysis which is neither a mindless ecology nor a glorification of mind divorced from the land" (Flannery and Marcus, 1976:383).

More recently, they have urged archaeologists to pursue a *holistic archaeology*, "a discipline in which cognitive variables would be given equal weight with ecological, economic and sociopolitical variables" (Marcus and Flannery, 1994:55). They observed that, although nearly all serious scholars recognize the importance of ideology and religion in the native cultures of the Americas, there is scant agreement about the methodology one should follow when researching these topics.

> The problem comes when we try to decide by which *scientific method* we will study these subjects. It is simply not enough to rely on one's intuition and assert what we believe to be true, as some of our humanistic colleagues have done of late. Cognitive archaeology needs a methodology just as "settlement and subsistence archaeology" does. (Marcus and Flannery, 1994:55)

Flannery and Marcus have proposed three major methods though which their holistic approach can be applied to specific archaeological situations: (1) the use of ethnographic and ethnohistoric data relevant to the archaeological situation under examination; (2) the analysis of public space, religious structures, and features; and (3) the contextual analysis of ceremonial artifacts (Marcus and Flannery, 1994:55).

In this chapter, we use such a methodology to demonstrate how a particular idea, ancestor veneration, was actualized through ritual in a small Middle/Late Formative (600–200 B.C.) village in the Cañada de Cuicatlán and thereby contributed to the legitimation of chiefly authority there. We then describe how the Cañada became subjugated by the expansionistic Monte Albán state during the Late/Terminal Formative (200 B.C.–A.D. 200). We show how this military takeover not only brought economic, social, and political changes to the Cañada, but also profoundly altered indigenous religious practices and beliefs.

THE CAÑADA DE CUICATLÁN

Ethnohistoric Background

The Cañada de Cuicatlán is a narrow *tierra caliente* canyon that serves as the only natural route between the Valleys of Oaxaca and Tehuacán (Figure 1). In the 16th century, the Cañada was occupied by a series of independent kingdoms, which the Spaniards called *cacicazgos*; the members of these groups spoke Cuicatec near

Figure 1. The Cañada de Cuicatlán, located between the Tehuacán and Oaxaca Valleys. (drawing by David Kiphuth)

Cuicatlán, Mazatec around Quiotepec, and Nahuatl-Popoloca in the Teotitlán del Camino area. All these *cacicazgos* were ruled by a dynastic elite, who traced their heritage back through a line of ancestors to a common, possibly mythical, progenitor named Teutli (Hunt, 1972:217, 254, Fig. 99). In each *cacicazgo* there was a head town and a number of subject settlements, all of which had leaders whom the Spaniards called *caciques*, although the cacique of the head town was often referred to as the *cacique mayor*.

Ethnohistoric evidence indicates that the ancestors of the caciques were venerated. The 1581 Relación de Teotitlán del Camino (Castañeda, 1905:213–231) contains a description of the ceremonies held in this *cacicazgo* throughout the year. The ninth fiesta in the cycle was called Hueymicailhuytl.

> La nouena fiesta llamauan *Hueymicailhuytl*; este dia tenian gran bayle los macehuales y los mandones y tequitatos de los barrios, y otros yndios particulares ofrecian mantas y cacao y maiz en la *tecpa* y casa del Cacique en honrra de sus difuntos, y la ofrenda era para el Cacique. (Castañeda, 1905:218–219)

The passage indicates that a great dance (*gran bayle*) was held, involving the commoners (*macehuales*) and members of the elite (*mandones* and *tequitatos*) and all other individuals. The participants brought offerings of cloth, drink, and food (*mantas*, cacao, and *maíz*) to the temple (*tecpa*) and house of the cacique in honor of his ancestors (*en honrra de sus difuntos*). Although we recognize the potential ambiguity in the use of the possessive pronoun in this last phrase, we think that a reading of the whole passage makes it reasonably clear that the ancestors honored by this ceremony were those of the cacique.

It thus appears that the ancestors of the cacique were venerated in a highly public context through communal ritual involving most if not all the subjects in the *cacicazgo*. We suspect that the ancestors of commoners were venerated in the context of household-level ritual, as Marcus has suggested for the Zapotec: "While chiefly ancestor worship may have been public, the worship of lower-ranking ancestors was part of household ritual" (Marcus, 1989:154; see also Flannery and Marcus, 1976:380–382).

Perdido Phase

Now let us shift our temporal perspective to the Middle/Late Formative period that we call the Perdido phase, 600–200 B.C., which is roughly contemporary with the Rosario phase and Monte Albán Ia in the Valley of Oaxaca. During our 1977–1978 fieldwork in the Cañada (Spencer and Redmond, 1997), we located 12 Perdido-phase sites, all situated on high alluvial terraces overlooking low alluvial fans (Figure 2). We have proposed that there were two *cacicazgos* in the Cañada at this time, each consisting of a head town and a number of subject villages (Redmond, 1983:62–73; Spencer, 1982:79–82). The polity to the south included sites Cs17, Cs22, Cs19, Cs21, and Cs25. The principal town of this group was

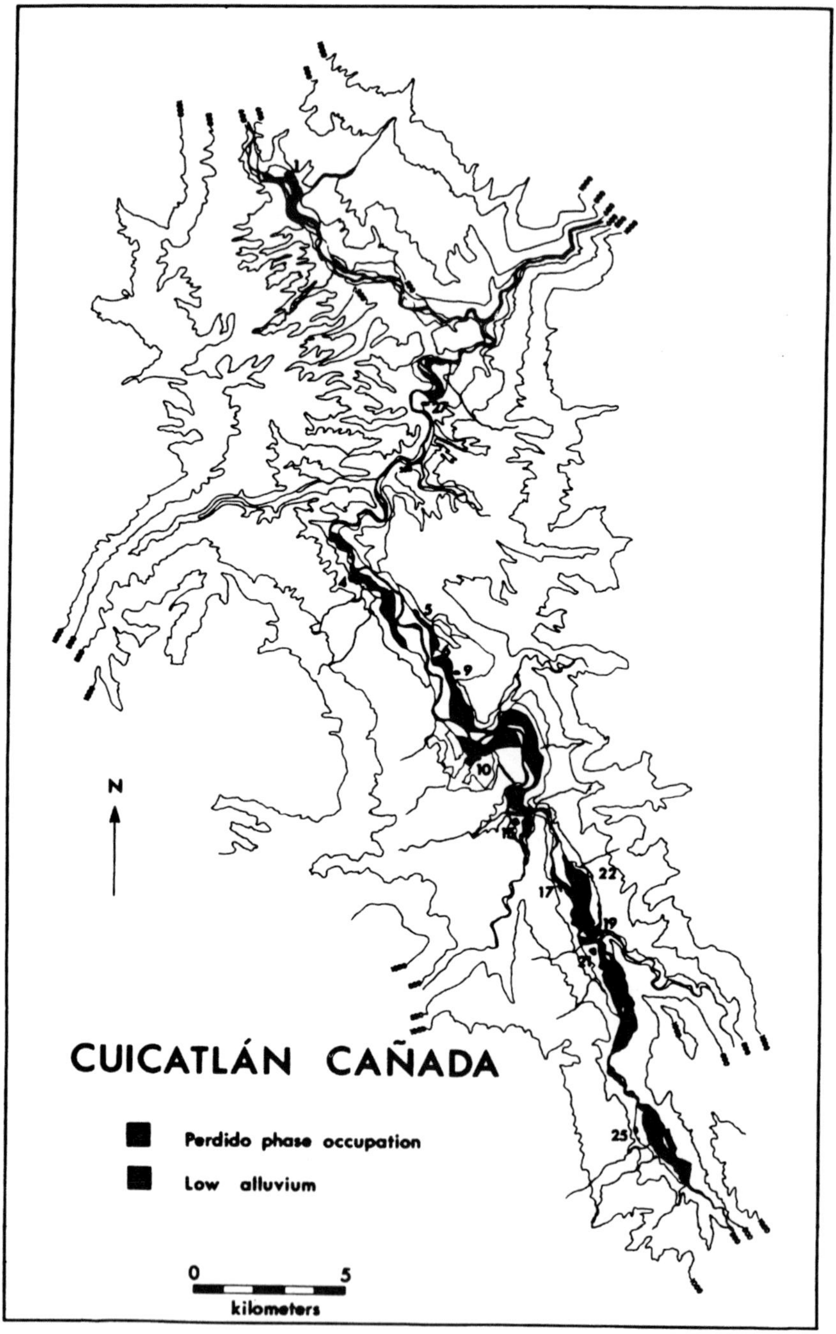

Figure 2. Perdido-phase regional settlement patterns in the Cañada de Cuicatlán.

Cs19. It covered 7.8 ha and had three mounds, up to 4 m in height, arranged around a plaza. The other sites were all less than 4 ha in area and had no visible mounds. We conducted excavations at one these sites, Cs25, or Llano Perdido.

Llano Perdido is situated on a high alluvial terrace overlooking the floodplain of the Río de las Vueltas. Our excavations here exposed over 1600 m^2 of this 2.25-ha village and showed that it was organized into four large residential compounds, each measuring about 30 m on a side. We were able to completely excavate one of these compounds, which we called Area A/B (Figure 3). It consisted of 18 structures arranged around three patios, including residences, storehouses, a tomb, and ceremonial platforms.

House 1 was a typical residence in Area A/B (Figure 4). It was evidently constructed by placing adobe bricks on the stone foundations and then supporting the roof with posts. In the southwestern corner of the house, we found a small stone-lined hearth (Feature 3).

On the western side of the Southwest Patio of Area A/B was House 7, the largest residence we found anywhere at the site (Figure 3). It was associated with three storage structures (H.8, H.9, and H.10), a tomb (St. 4), and a ceremonial

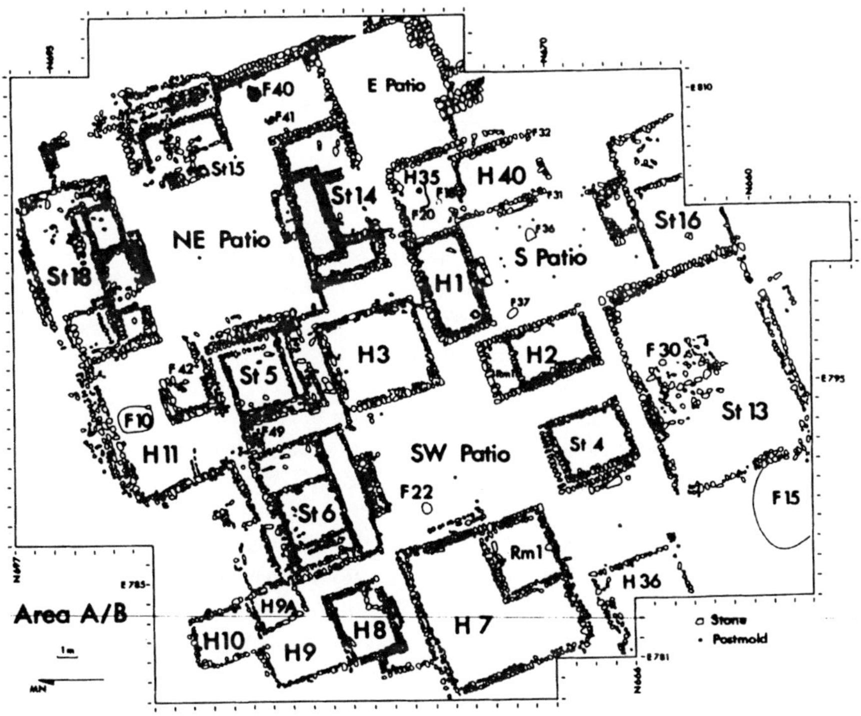

Figure 3. Plan of Area A/B at Llano Perdido.

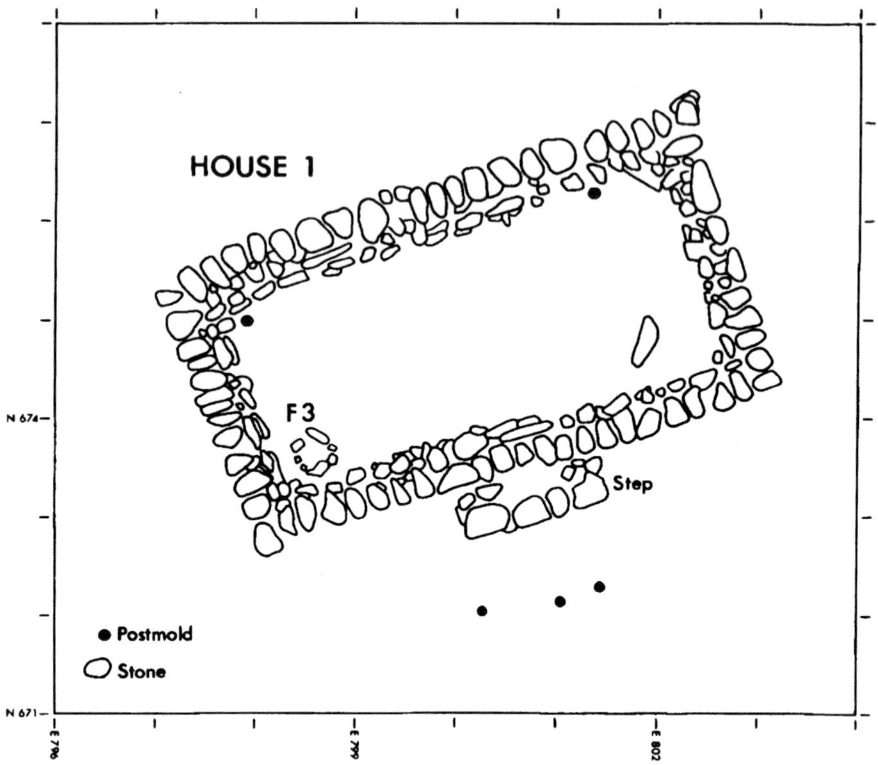

Figure 4. Plan of House 1 in Area A/B.

platform (St. 6). In Figure 5, House 7 can be seen in the foreground. Structure 4, the tomb, is on the right, while on the left is Structure 6, with a person sitting on the steps leading to the top of the platform.

The Structure 4 tomb contained the most elaborate burial that we excavated at Llano Perdido (Figure 6). This was Burial 2, an adult male interred with 30 ceramic pots and a dozen ornaments of marine shell. Among the shell ornaments, biologist Oscar Polaco identified a smoothed, polished bivalve half as *Patella mexicana* and nine drilled shells (some with traces of red paint) as *Pyrena fuscata*, all from the Pacific Ocean (Polaco, 1997:612–613). Physical anthropologist Richard Wilkinson observed that one of the individual's upper incisors had been filed, and he noted some red pigment on the parietal bone (Wilkinson, 1997:614). These features and the large number of accompaniments suggest that the deceased was a very high-status male. We suspect, in fact, that he may have been the highest-ranking person at Llano Perdido, none other than the cacique of the village.

The most frequent kind of vessel in the tomb was a Perdido Gray outleaned-wall serving bowl with incised decoration. Eight of the thirty vessels accompany-

Figure 5. View of Area A/B, looking east, with House 7 in the foreground, the Structure 4 tomb to the right, and the Structure 6 platform to the left.

ing Burial 2 were of this type. A representative example is Vessel 24, shown in Figure 7. Although several motifs were incised on these vessels, it is noteworthy that *all* eight outleaned-wall bowls have a distinctive, repeating steplike motif on their interior rims (see Figure 7), which we called Motif 17. We noted that this motif also appeared on four other vessels (two ovate-rim bowls, a small chalice, and a *vaso*, all of Perdido Gray)—making a total of 12 whole vessels with Motif 17 in Structure 4. Given its prevalence in a high-status tomb, we were convinced that this motif was of considerable importance at Llano Perdido. But just what did it signify? As we researched this question, we became intrigued by the similarity between Motif 17 and what Marcus interpreted as the "Lightning" motif on Vessels 1 and 3 from Burial 1 at the Formative site of Abasolo in the Valley of Oaxaca (Marcus, 1989:Figs. 8.12, 8.13). A reasonable hypothesis, we now propose, is that Motif 17 at Llano Perdido is also a depiction of "Lightning." Let us consider some of the implications of this hypothesis.

We agree with Marcus that the Lightning motif probably had important supernatural associations. Ethnographic data from the traditional Cuicatec town of San Andrés Teotilalpan indicate that Lightning (or Rayo) was the most important supernatural entity; it was believed to control the rain (Hunt, 1972:207). In the Valley of Oaxaca, Marcus and Flannery have reported that the elite claimed descent from the supernatural force Lightning, whom the Zapotec named Cociyo

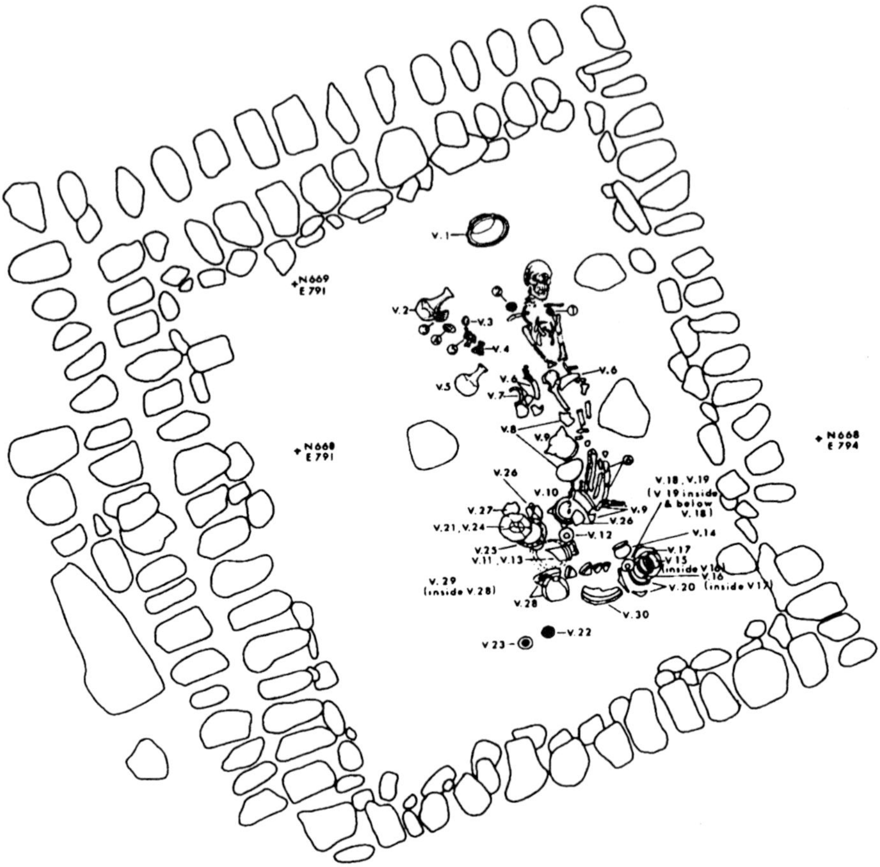

Figure 6. Plan of Burial 2 in Structure 4 in Area A/B.

(Flannery and Marcus, 1976; Marcus, 1978, 1989; Marcus and Flannery, 1978, 1994). Marcus (1989:150–151) pointed out that Cuicatec and Zapotec belong to the Otomanguean language family along with Mixtec, Mazatec, Popoloca, Chatino, Mixe, and others; most of these cultures revered Lightning as the most powerful supernatural force (see also Marcus et al., 1983:38).

Because the Zapotec were the Cañada's nearest neighbors to the south, Flannery and Marcus's findings about Zapotec ideology are surely germane to the present discussion. They reported that the Zapotec of the 16th century conceived their quadripartite world as pulsating with a variety of natural and supernatural forces, all of which possessed a vital force or spirit (*pèe*) that animated the earth, living things, and natural forces such as wind, thunder, lightning, and earthquakes. The spirit in Lightning (Cociyo) was a dominant force because it caused thunder and rain; consequently, it was greatly revered by the Zapotec (Flannery and Marcus,

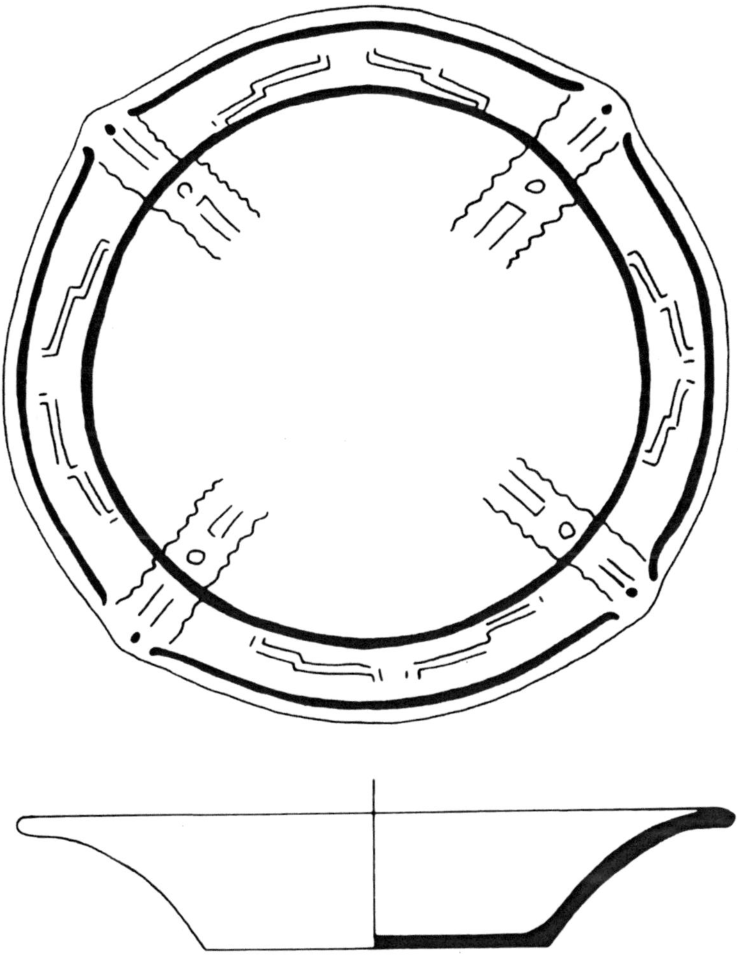

Figure 7. Perdido Gray serving bowl with incised decoration (Vessel 24 from Burial 2, Area A/B, Llano Perdido).

1976:376, Marcus, 1989:150–151). The Zapotec thought that Cociyo split the clouds, causing thunder (Xòo Cociyo, literally "Lightning's Earthquake") to roll and making the rain fall (Marcus and Flannery, 1978:58, 1996:95). To ask for more rain, the Zapotec offered Cociyo blood, drawn from their own bodies with sting-ray spines or obsidian blades in self-sacrficial rituals, or the beating hearts of sacrificial victims, which still possessed *pèe* (Flannery and Marcus, 1976).

The spirits of dead ancestors joined the supernatural forces and, if respectfully honored through rituals and offerings, could intercede on behalf of their living descendants. Marcus (1989) reinterpreted certain ceramic motifs incised on Formative Oaxaca pottery in light of this ideology of descent practiced by Formative Oaxaca villagers. She proposed that the bold, angled lightning bolts (often in pairs) that appear in what some have called the fire-serpent motif can be interpreted as a vivid depiction of Lightning, whereas the open cleft in the bracketed frame of what has been termed the were-jaguar motif may actually represent a fissure or earthquake emanating from a quadripartite earth. In the Early Formative period, these motifs were strongly associated with houses and male burials in separate residential barrios at San José Mogote, and it has been suggested that they probably served as the patron ancestors of separate male descent groups (Marcus and Flannery, 1996:95–96; Pyne, 1976).

During the Middle Formative period, when ranked societies were well established in the Oaxaca Valley, the Lightning motif persisted in the form of a stylized double-line break motif in the region's ceramics (Flannery et al., 1981:76–78; Marcus, 1989:194–196). Marcus suggested that the Zapotec elite asserted its claimed descent from Cociyo through villagewide ceremonies that the elite directed (Marcus, 1989:154). Later, the Cociyo theme appeared on Classic-period funerary urns, thought by Marcus (1983) to represent the venerated ancestors of the elite. As Marcus (1989:196) has noted: "Once the ranking of Formative times had evolved into the stratification of Classic times, 'lightning' (Cocijo) emerged as the most powerful supernatural associated with the professional ruling class at Monte Albán."

In view of these data and interpretations, it is perhaps not surprising to observe a Lightning motif on Perdido Gray bowls in the high-status Structure 4 tomb at Llano Perdido. It is also notable that the Lightning motif is usually accompanied on these vessels by another motif consisting of wiggly vertical lines (Figure 7); we interpret this as a depiction of rain. Still another motif that usually appears on these vessels is the double-line break (Figure 7). We propose that this motif is related to (and perhaps descended from) the open cleft motif that Marcus (1989) interpreted as Earthquake; when associated with Lightning, this motif likely represents Lightning's Earthquake or thunder. We consequently suggest that these vessels in the Structure 4 tomb at Llano Perdido display an iconographic association among Lightning, thunder, and rain. The vessels were probably used in the funerary rites that accompanied the interment of the village cacique. As we shall see, however, they were used in certain other contexts as well.

Across the Southwest Patio from the tomb, and adjacent to House 7, lay Structure 6 (Figure 3). This was a platform, constructed in the manner often seen in Mesoamerica, with a network of interior reinforcing walls. This structure was associated with a high density of ceremonial artifacts such as obsidian blades (which may have been used in autosacrificial bloodletting), figurines, red pigment, and a deer antler rattle (Spencer 1982:111–114).

The cacique was apparently quite well off in a material sense. Not only was his residence (House 7) the largest in the entire village, but adjacent to it was the greatest concentration of storage facilities found at the site—Houses 8, 9, 9A, and 10. These had stone foundations, but also contained high densities of stick-impressed daub, indicating that they had walls of wattle and daub. Unlike the Valley of Oaxaca, where bell-shaped storage pits are common, the Cañada's inhabitants apparently preferred aboveground storage. In fact, Houses 8, 9, 9A, and 10 are similar in appearance to the storehouses of the ethnographically known Cuicatec in the region (Weitlaner 1969:436–439, Fig. 3). These structures at Llano Perdido had high relative frequencies of imported pottery, obsidian, and shell—all of them rare, high-cost items that can be reasonably interpreted as prestige goods or symbols of wealth (Spencer, 1982:114–118, Table 3.4).

In sum, the Southwest Patio was the residential area of what we think was the village cacique, who lived in the village's largest residence (House 7), had access to more stored wealth items than anyone else in the community (Houses 8, 9, 9A, 10), who conducted certain ritual activities atop the Structure 6 platform adjacent to his house to the north, and who was destined to be buried in the Structure 4 tomb.

About 25 m northeast of Area A/B was Area C, another residential compound at Llano Perdido. We did not have enough time to excavate all of Area C, but we did find another residence, platform, and tomb, in a layout like the one just described for Area A/B. This is House 20 (residence), Structure 17 (platform), and Structure 12 (tomb). Although the tomb in this case had only three ceramic vessels and no shell ornaments, the single gray outleaned-wall serving bowl was inscribed with the Lightning motif, just like those in the Structure 4 tomb. The individual buried here was also an adult male, and he was probably the highest-ranking member of the Area C compound, although he was undoubtedly of lower status than the individual buried in the Structure 4 tomb in Area A/B.

We have proposed that the Llano Perdido community was organized into four residential wards, each of which was directed by a headman (Spencer, 1982:131–137, Fig. 3.27). One of the wards (Area A/B) was ranked higher than the others, and the head of the highest ranking ward also served as the cacique of the village. As did the caciques of the 16th century, he probably traced his heritage back to real or mythical ancestral chiefs.

Ordinary households also carried out ritual activities; the presence of figurines and obsidian blades in most of the residences indicates that household-level rites did take place regularly, along with the usual domestic activities such as maize grinding (Spencer, 1982:Tables 3.2, 3.3). Yet, it appears that there were also ceremonies that occurred at Llano Perdido on a suprahousehold, communal basis. Some of these may have taken place in the Northeast Patio. The Northeast Patio (Figure 3) had platform structures on all four sides; grinding stones were completely absent here, although ceremonial artifacts were abundant (Spencer, 1982:118–124). It is noteworthy that no such patio group was found in Area C, the other excavated compound. In fact, we suspect that the Northeast Patio was the main public space for the entire Llano Perdido communty, and it was directly associated with what we think was the highest-ranking lineage and its leader.

One activity that probably occurred in the Northeast Patio with some frequency was communal feasting. Just off the northwest corner of the patio was House 11, a specialized cooking structure with a very large hearth (Feature 10). Feature 10 is approximately 28 times larger in area than the Feature 3 hearth inside House 1; we consequently suggest it could have been used to prepare a meal for the members of 28 such residences. Because an independent estimate of the number of residences at the Llano Perdido site ranges from 24 to 33 (Spencer, 1982:130), it seems likely that the entire village could have participated in a ceremonial feast prepared here. Significantly, one of the Northeast Patio platforms, Structure 5, is directly connected to Structure 6, the platform associated with what we think was the cacique's residential patio (Figure 3). The cacique could have easily passed from Structure 6 to Structure 5, where he could have directed any ceremonies taking place in the Northeast Patio.

We noted earlier that the Lightning motif occurred on Perdido Gray bowls that accompanied two elite burials in tombs (Structure 4 and Structure 12) at Llano Perdido. In what other contexts at the site do we find vessels with this motif? In Figure 8 we present a bar graph showing the distribution of Perdido

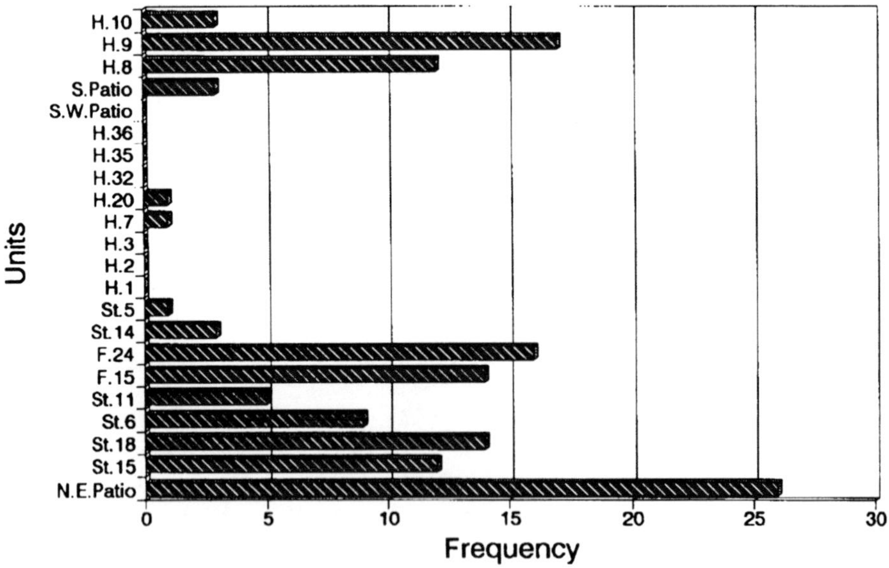

Figure 8. Bar graph showing the distribution of the Lightning motif among excavated cultural units at Llano Perdido.

Gray bowl sherds with the incised Lightning motif in 22 excavated cultural units from Llano Perdido. We should stress that all these units have been dated securely to the Perdido phase. The highest frequency of this motif occurs in the Northeast Patio, which is where we have suggested communal feasts were held (Figure 8). The motif is also present at all the platform structures around the Northeast Patio (Structures 18, 15, 14, 5, and 11), as well as in a dump just north of this patio (Feature 24). The motif appears as well in Feature 15, a dump associated with Structure 13, possibly also a ceremonial construction. In addition, the Lightning motif is well represented at Structure 6, the ceremonial platform connected to the Northeast Patio and associated with House 7, the probable cacique's house. Houses 8, 9, and 10, the storehouses associated with House 7, also have high frequencies of sherds with the Lightning motif.

On the other hand, the motif was completely absent from House 1, the ordinary residence that we described earlier, and also from Houses 2, 3, 35, and 36, all in Area A/B; nor does it appear in House 32 in Area C (Figure 8). The Lightning motif appeared on a single sherd in each of only two residences: House 7, the likely residence of the village cacique, and House 20, the probable house of the Area C headman. The elite occupants of these two households may have used the bowls with Lightning motifs for their own meals. Given the very sparse occurrence in Houses 7 and 20, however, such feasts were probably much less common than those that evidently occurred in the Northeast Patio area.

Table 1 contains a list of the 22 excavated proveniences, together with their inferred cultural contexts. If we combine this information with that from Structures 4 and 12, it is clear that the Lightning motif occurred at Llano Perdido in the following contexts: (1) elite tombs; (2) elite residences; (3) elite storage; (4) elite ceremonial structures adjacent to elite residential areas; and (5) public ceremonial structures and spaces. The complete absence of the motif in nonelite residences is striking. We suggest that ordinary villagers did not use vessels with the Lightning motif in their own homes, but only when they were participating in public ceremonies, such as the large-scale communal feasts that we have suggested took place in the Northeast Patio.

The interment of a village cacique in his community was surely a symbol of his continuing importance in village affairs. This burial location, along with the vessels and other offerings placed in his tomb, can be reasonably understood as an expression of elite ancestor veneration. The distribution of the Lightning motif among the other cultural units, however, is perhaps telling us something more profound. What we think we have here is evidence of an important *ideological linkage* expressed through ritual at Llano Perdido. This linkage began with the ~~funerary ritual and tombs of deceased leaders, continued through the households~~ of the living leaders, and ultimately connected with public ceremonies that occurred in a suprahousehold, probably villagewide, context. In sum, a bond existed between the ancestral chiefs and ordinary villagers—a bond that was sanctified

Table 1. Perdido Phase Excavated Units with Their Inferred Cultural Contexts

Unit	Contexts
N.E. Patio	Ceremonial
St. 15	Ceremonial
St. 18	Ceremonial
St. 6	Ceremonial
St. 11	Ceremonial
F. 15	Ceremonial
F. 24	Ceremonial
St. 14	Ceremonial
St. 5	Ceremonial
H. 1	Domestic
H. 2	Domestic
H. 3	Domestic
H. 7	Domestic
H. 20	Domestic
H. 32	Domestic
H. 35	Domestic
H. 36	Domestic
S.W. Patio	Domestic
S. Patio	Domestic
H. 8	Elite storage
H. 9	Elite storage
H. 10	Elite storage

by Lightning, mediated by the living cacique, and continually reaffirmed through public ceremony.

Lomas Phase

Our tale does not end here. Major changes were brewing in the Oaxaca area during the Late/Terminal Formative period (Monte Albán Ic–II phases, c. 300 B.C.–A.D. 200). The epicenter of many of these changes was in the Valley of Oaxaca itself, at the great site of Monte Albán. During this time, Monte Albán grew to become a city with an estimated population of about 15,000 persons living on its hillside terraces. By Monte Albán II (100 B.C.–A.D. 200), the Main Plaza of Monte Albán had been laid out. Almost two dozen public buildings were constructed here, including the arrowhead-shaped structure called Building J, standard two-room temples, the ball court, and an outdoor adoratory. A 3-km-long defensive wall was erected along the gradual northern and western limits of Monte Albán. Furthermore, on the regional level, Monte Albán stood at the top of a four-tiered settlement-size hierarchy in the Valley of Oaxaca, made up of secondary administrative centers, tertiary centers, and villages. The settlement-size hierarchy and the abundance and variety of public architecture constructed at Monte Albán are compelling evidence that the Zapotec state had formed by the beginning of Monte Albán II (Marcus and Flannery, 1996:172–194).

On Building J were some 50 carved stone slabs whose inscriptions, according to Caso (1947) and Marcus (1976, 1980), commemorated conquered places. One of the specific places listed as having been conquered is represented by a human head with a feathered speech scroll, which is similar to the glyph for Cuicatlán (or "Place of Song") in the Codex Mendoza. Based on these inscriptions, Marcus hypothesized in 1976 that Monte Albán launched a campaign of conquests in the Late/Terminal Formative, directed at, among other regions, the Cañada de Cuicatlán. The data that we recovered in the Cañada in 1977–1978 are consistent with the conquest hypothesis (Redmond, 1983; Spencer, 1982; Spencer and Redmond, 1997). It appears that the Cañada was subjugated by the Monte Albán state at the interface between the Perdido phase (600–200 B.C.) and the Lomas phase (200 B.C.–A.D. 200) and remained in a subordinate, probably tributary, relation throughout the Lomas phase.

A major settlement-pattern disruption occurred at the beginning of the Lomas phase: All Perdido-phase sites were abandoned, and new settlements were founded on nearby slopes and ridges (Figure 9). In the Quiotepec area at the northern end of the Cañada, the single small Perdido-phase site was succeeded by a 45-ha complex of seven sites that sprawled across both sides of the natural pass into the Cañada from Tehuacán as well as the strategic mountain peaks (Figure 10). Heavily fortified, the Quiotepec sites were clearly a military frontier installation, designed to control movement through the northern frontier of the Cañada (Redmond, 1983:91–120). The Quiotepec installation also marks the northern limit of Lomas-phase pottery, some of which is similar to the pottery of the contemporary Monte Albán Ic and II phases in the Valley of Oaxaca (Redmond, 1983:86). We think this ceramic boundary resulted from a deliberate strategy by Zapotec frontier administrators to prevent uncontrolled contact between the conquered inhabitants of the Cañada and the people of the Tehuacán Valley to the north.

Part of the frontier complex was Cerro de Quiotepec (Cs2), a massive hilltop fortress with huge mounds and defensive walls (Figure 11). Flanking the natural pass was Paso de Quiotepec (Cs3), a 15-ha site with an enormous I-shaped ball court, mounds, and walls (Figure 12). Immediately across the river from the pass, at a ford that is still in use, was Campo del Panteón (Cs26), a large (9 ha) plaza with impressive mounds; anyone traveling into the Cañada from the north would have had to transit this plaza before proceeding south (Figure 13).

We have estimated that a total of 443–2,050 people lived in the Quiotepec locality during the Lomas phase, a much higher population than we observed for any other period, including the present (Redmond, 1983:96). The sustenance of this extraordinary occupation almost certainly required that food be brought from outside the Quiotepec locality. Our calculations suggest that there were more than

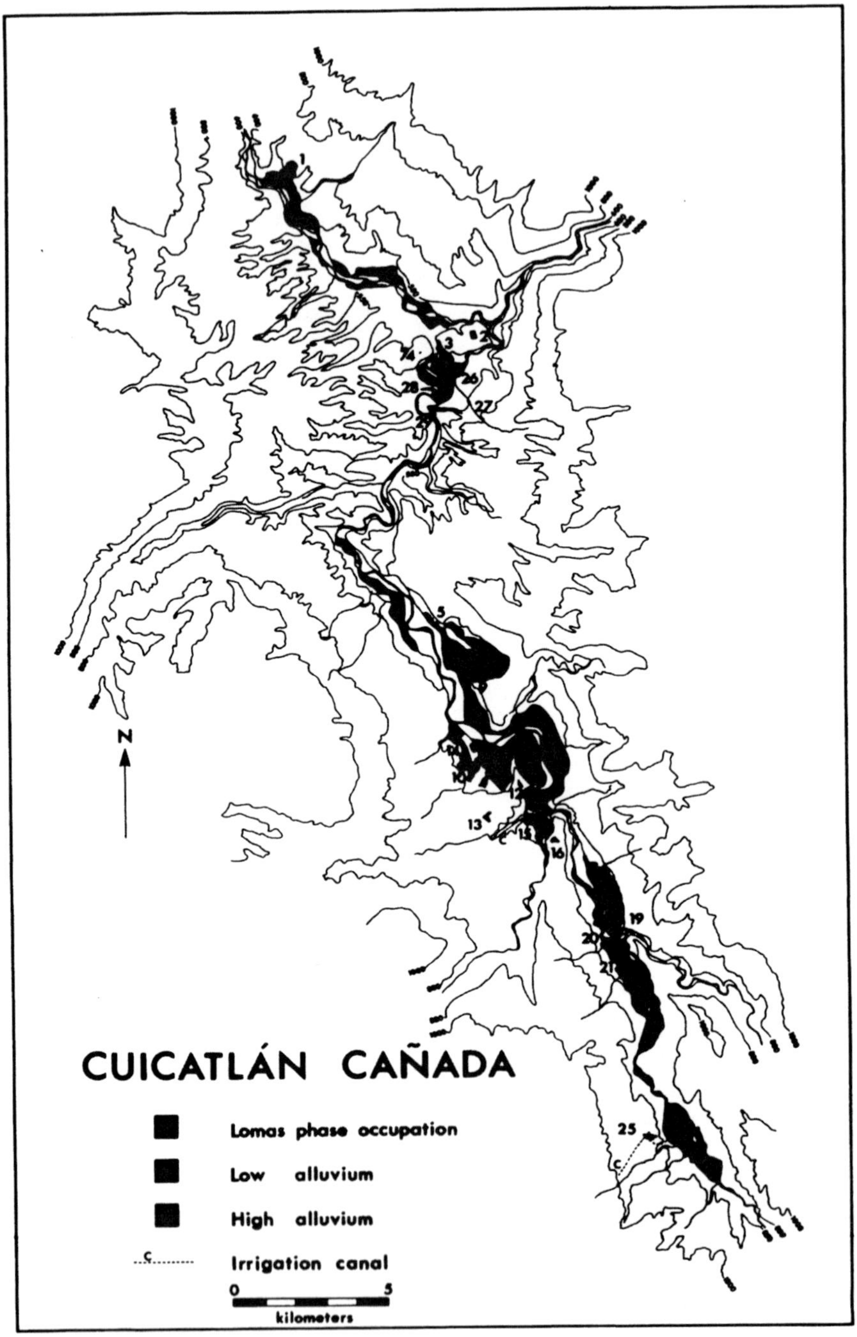

Figure 9. Lomas-phase settlement patterns in the Cañada de Cuicatlán.

Figure 10. View of the Quiotepec locality, facing north, showing the mountain pass that is the only natural route into the Cañada from the Tehuacán Valley. The site of Paso de Quiotepec (Cs3) occupied both sides of the pass itself; just across a ford on the Río Grande lies Campo del Panteón (Cs26), whose plaza with tree-lined mounds can be seen in the lower center.

Figure 11. View of the hilltop fortress of Cerro de Quiotepec (Cs2), looking northwest.

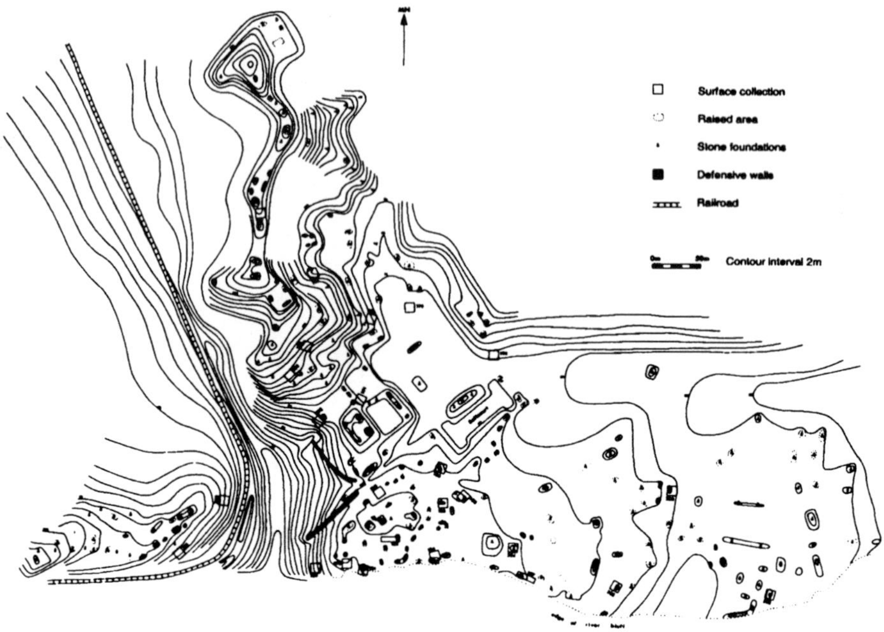

Figure 12. Map of Paso de Quiotepec (Cs3).

four times as many people residing in this locality during the Lomas phase as could have been supported by farming the arable land in just the Quiotepec locality (Redmond, 1983:105–106). We suspect they were sustained by surplus mobilized from the other alluvial fans of the Cañada, the nearest of which lay just 7 km south of Quiotepec.

In the central and southern parts of the Cañada, there were also great changes in the Lomas phase. The village of Llano Perdido was abandoned at this time. In fact, it was burned to the ground, and we found the body of a woman lying on an interior house floor, covered by the collapsed walls and debris of her home (Spencer, 1982:216–220). In this abrupt and evidently violent fashion, settlement shifted to Loma de La Coyotera, a nearby ridge.

Covering just 3.04 ha, the Lomas-phase settlement on the Loma de La Coyotera ridge was only slightly larger than the 2.25-ha Perdido-phase community on the high alluvium (Figure 14). Nevertheless, the Lomas-phase village differed from its predecessor in a number of ways. The basic unit of residence changed from the large compound of earlier times to a pattern of one or two residences on a terrace carved into the hillside. We suspect that the Zapotec frontier administrators may have pursued a deliberate policy of breaking up the traditional residential compounds to disrupt interfamilial relationships; this would have allowed for

Figure 13. View of Campo del Panteón (Cs26), looking south.

more effective control over the conquered Cañada population. Like the other 13 settlements located on hilltops in the central and southern Cañada, the public sector of Loma de La Coyotera consisted of a single plaza, in this case with two small mounds on a 3-m-tall platform. At Loma de La Coyotera, we located 35–38 residential terraces and a total of 55–60 house mounds or house foundations.

We also found an elaborate irrigation facility at Loma de La Coyotera. We mapped the route of an ancient canal that brought water down from a tributary *barranca* behind the site, over a dozen gullies and depressions where aqueducts were raised, and channeled it onto the expanse of high alluvium where the previous Perdido-phase community of Llano Perdido had been located (Figure 14). Our excavation of the largest aqueduct revealed that this facility was first constructed in the Lomas phase, when the high alluvium was abandoned in favor of the ridgetop.

With the introduction of sophisticated canal irrigation, the Cañada's 740 ha of high alluvium could have been brought under cultivation (joining the 997 ha of low alluvium), resulting in an enormous expansion of the agricultural regime. Because the local population residing at settlements in the central and southern Cañada remained far below the carrying capacity of just the low alluvium (Redmond, 1983:123–126), it is likely that Lomas-phase production was expanded largely in response to demands by the Zapotec state for tribute. The items produced as tribute probably included basic foodstuffs, particularly important for the

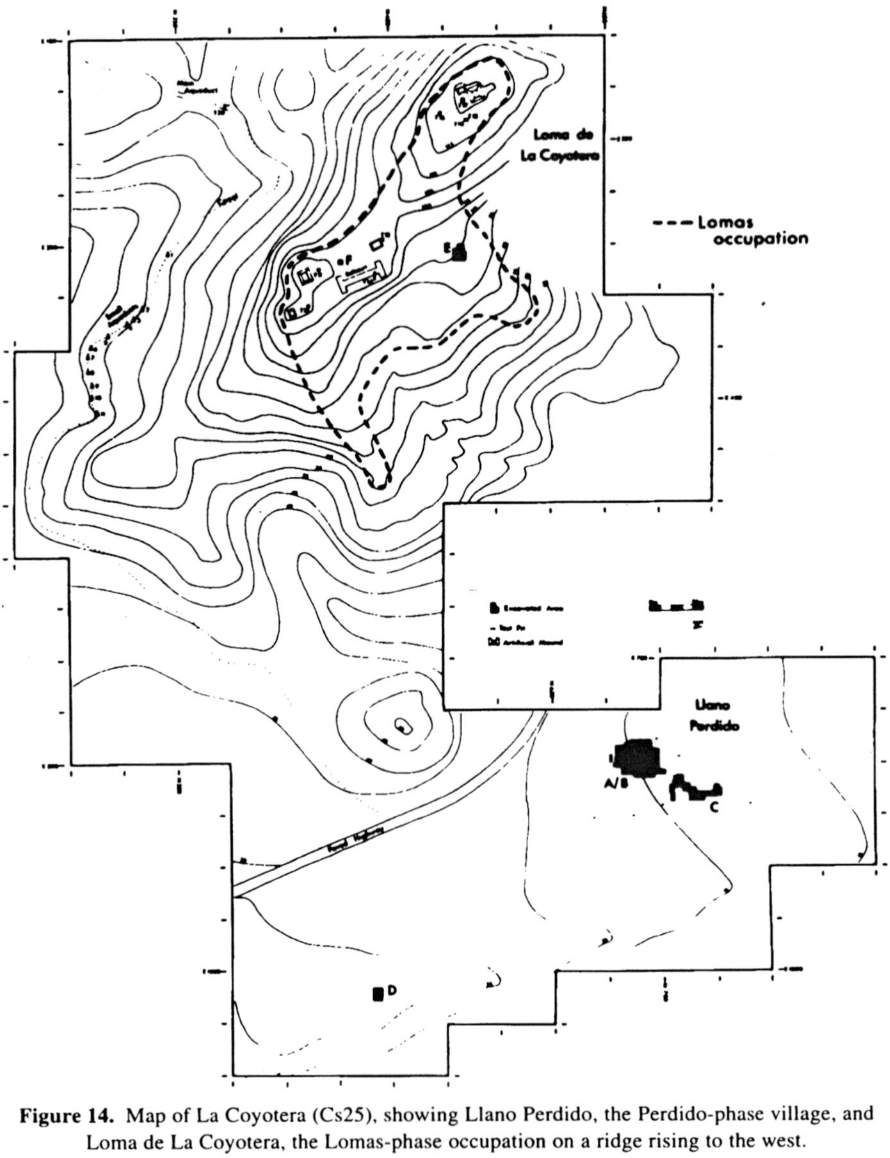

Figure 14. Map of La Coyotera (Cs25), showing Llano Perdido, the Perdido-phase village, and Loma de La Coyotera, the Lomas-phase occupation on a ridge rising to the west.

provisioning of the frontier facility at Quiotepec. Yet there is also evidence that certain tropical fruits and nuts were being cultivated here on a large scale, possibly in response to tribute demands. We excavated a Lomas-phase midden deposit at Loma de La Coyotera that contained extremely high densities of tropical fruit and nut remains, especially the *coyol* palm (*Acrocomia mexicana*), the black *zapote*

(*Diospyros digyna*), and the *ciruela* (*Spondias mombin*). These items occurred in much higher densities than was the case in the earlier Perdido-phase midden deposits at the Llano Perdido site (Spencer, 1982:227–230). Thus, it is likely that much of the newly irrigated high alluvium was sown with these *tierra caliente* species, all of which require considerable amounts of water. With the increase in the production of these plants at La Coyotera came a decrease in the variety of craft activities performed. Our Lomas-phase deposits reveal a sharp drop in the relative frequencies of craft-related artifacts (such as spindle whorls, sherd disks, and loom weights) from Perdido times (Spencer, 1982:230–231). This pattern may reflect a policy of enforced economic specialization in the central and southern Cañada, enacted by Zapotec administrators to increase the relative labor time spent in tribute production.

In the Lomas phase, the inhabitants of tributary communities such as La Coyotera evidently had little access to luxury goods imported from other regions. Our surface collections in the central and southern Cañada (south of the Quiotepec frontier sites) show a marked reduction from the Perdido to the Lomas phase in the relative number of fancy Oaxacan ceramics, obsidian, and marine shell (Redmond and Spencer, 1982:30).

Ritual practices in the Cañada were also dramatically transformed between the Perdido and Lomas phases. At La Coyotera, the ceremonial platforms and courtyards of the Perdido-phase community were replaced by a single plaza with two pyramidal mounds at the Lomas-phase occupation. Our excavated Lomas-phase deposits contain far fewer ceremonial artifacts, such as figurines and obsidian blades, than our Perdido-phase samples (Spencer, 1982:Table 5.7). It is also noteworthy that Lightning motifs are almost never found on Lomas Gray bowls. In fact, only one sherd with the Lightning motif was excavated at Loma de La Coyotera; it was associated with House 24, a Lomas-phase residence in Area E (Spencer and Redmond, 1997:Table 9.2). We conclude that the rich ceremonial life evident at the earlier community of Llano Perdido—replete with costuming, music making, communal feasting, and self-sacrificial bloodletting—had virtually disappeared by the Lomas phase.

Instead, our excavations in the plaza of the hilltop Lomas-phase community recovered an awesome display of imposed authority. In front of the main pyramid mound, we exposed the remains of at least 61 human skulls, distributed in rough rows, along with numerous scattered skull fragments and at least one post mold. We have proposed that this feature represents an overturned skull rack that had consisted of upright posts with skulls mounted on crossbeams (Figure 15). Skull racks (or *tzompantli*) have long been associated with the militaristic Toltec and Aztec of the later Postclassic period, who used them to display the heads of captives taken in warfare. From the writings of the Spanish chronicler Francisco de Burgoa (Burgoa, 1674b:342), we know that the Postclassic Zapotec also constructed skull racks and that they did so in conquered regions when they wanted to

Figure 15. Artist's reconstruction of the *yàgabetoo* or skull rack at Loma de La Coyotera. (drawing by Diana Salles)

terrorize the peoples they had subjugated. Marcus and Flannery (1996:205–206) reported that the Zapotec word for skull rack was *yàgabetoo*. Because the skull rack at Loma de La Coyotera was associated with clear Lomas-phase diagnostic ceramics (Spencer and Redmond, 1997:522–523), we can now trace this terror tactic back to the end of the Formative period.

On the same mound platform where the skull rack stood, we recovered the paw from a large, hollow ceramic jaguar statue (Figure 16). A similar jaguar paw fragment also turned up in a surface collection in the plaza of another Lomas-phase hilltop site in the Cañada. They are both virtually identical to the paws on the jaguar statue or urn recovered at the Monte Albán II secondary center of Suchilquitongo at the northern end of theValley of Oaxaca (Figure 17). A similar jaguar statue, we suspect, stood alongside the *yàgabetoo* at Loma de La Coyotera. Let us consider the possible significance of such jaguar imagery.

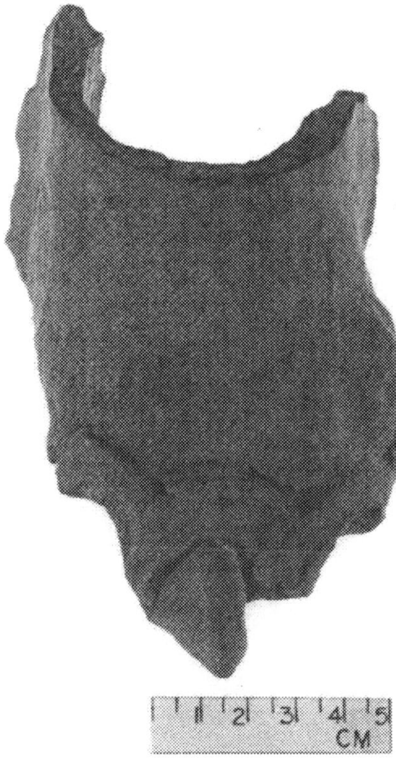

Figure 16. Fragment of hollow ceramic jaguar paw found near the *yàgabetoo* at Loma de La Coyotera (scale in cm).

When the Zapotec conquered the Cañada de Cuicatlán in the Late Formative, they transformed it into a frontier, tributary region of the Zapotec state centered at Monte Albán. We suggest that, as part of this process, native rule by sanctified chiefs in the Cañada was replaced by a new imperial ideology that emanated directly from Monte Albán. What were the major elements of this imperial ideology? We should note that Caso, Marcus, and other scholars have suggested that Monte Albán's ancient name may have been "Hill of the Jaguar" (Caso et al., 1967:84; Marcus, 1976:131, 1980:58). The magnificent ceramic jaguar statue shown in Figure 18, now on exhibit at the National Museum of Anthropology in Mexico City, was found in a Monte Albán II deposit in the West Platform, bordering the Main Plaza of Monte Albán (Caso and Bernal, 1952:55). Caso and Bernal pointed out that jaguar statues, urns, and figurines (what they called *representaciones de tigres*) became much more common at Monte Albán beginning in Monte Albán II.

Figure 17. Ceramic jaguar statue or urn (actual height 43 cm) from the Monte Albán II site of Suchilquitongo in the Valley of Oaxaca (redrawn from Caso and Bernal [1952:63]). (drawing by David Kiphuth)

> Las representaciones de tigres concebidas como un animal, y no ya con características humanas, son bastante abundantes. La más antigua encontrada en Monte Albán es la de un gran tigre de la época II, que apareció en el basurero de la Plataforma Oeste. . . . A partir de esta época, las representaciones de tigres son muy frecuentes en Monte Albán. (Caso and Bernal, 1952:55)

The representations of jaguars proliferated at Monte Albán, in the form of freestanding jaguar statues, jaguar heads, jaguar attributes (such as paws and fangs) on vases, urns, and even on the decorative panels of temple platforms at Monte Albán (Caso and Bernal, 1952:28–33, 54–64). Marcus (1976:133) has indicated that jaguars, and men depicted as jaguars, are well represented in the carved stones at a number of other Monte Albán Ic–II sites in the Valley of Oaxaca.

As time went on, Zapotec rulers at Monte Albán were often shown with jaguar attributes. A well-known example can be seen on the Lápida de Bazán, a Classic-period stone slab that portrays an individual known as 3 Turquoise, called "the god of Monte Albán" by Caso and Bernal (1952:54). In this inscription, the

Figure 18. Ceramic jaguar statue (actual height 85 cm) from a Monte Albán II deposit in the West Plataform, associated with the Main Plaza of Monte Albán (Caso and Bernal, 1952:55). Photographed by the senior author at the National Museum of Anthropology, Mexico City.

figure on the right is shown with an elaborate headdress and with jaguar paws instead of human hands or feet (Marcus and Flannery, 1996:Fig. 281). Marcus (1980:62) has suggested that 3 Turquoise was an actual human, a ruler or some other high official of Monte Albán, who is shown in the inscription as a "jaguar lord." It is likely that the power and authority of Monte Albán's rulers were based, in part, on their military prowess. We also think that the jaguar became a principal symbol of the Zapotec state itself.

As Marcus (1989) pointed out, the Lightning (or Cociyo) theme continued to be associated with the social (probably endogamous) class of Zapotec elite. However, not all members of this elite were depicted with jaguar attributes. We suggest that only those who held important administrative or military positions in the Zapotec state were shown in the inscriptions as jaguar lords. Perhaps it was their *political role*, not simply their elite status in Zapotec society, that determined whether certain individuals would be identified with the jaguar, symbol of the state.

We propose that when the Zapotec embarked on the conquest of neighboring regions, they took this symbolism with them. As the coercive hand of the Zapotec state intruded into the affairs of conquered regions, the jaguar became the new overlord of the subjugated peoples, supplanting the local leaders. We suspect that the Zapotec state may have purposefully discouraged the use of the Lightning motif in conquered villages as part of a general strategy aimed at undermining the traditional, sanctified authority of local leaders. In the case of the Cañada de Cuicatlán, this period of domination lasted until about A.D. 200, after which the local Cañada caciques began to reassert their authority—under the watchful eye of the ancestors, and of Lightning.

CONCLUSION

We would like to underscore two points by way of conclusion. A good bit of the analysis in this chapter involved a consideration of the *cultural contexts* of ceremonial artifacts and features. For example, we addressed the religious and sociopolitical significance of the Lightning motif by examining the *distribution* of sherds with that motif among residences of varying status, storehouses, tombs, and ceremonial plazas and structures at the Llano Perdido site. In doing so, we followed the methodology espoused by Kent Flannery, who for years has urged the archaeologist "to look to the contexts of his ritual or religious discoveries for clues to their function in the overall system" (1976:331). It is worth emphasizing that to carry out such a contextual analysis we needed *exact provenience data* on a whole series of structures, features, artifacts, and specific design elements. Moreover, to help us interpret the patterns we observed in the archaeological record, we drew on ethnohistoric and ethnographic data from Oaxaca. This analytical blend of archaeology, ethnohistory, and ethnography was also inspired by Flannery's research. Second, although we paid much attention to iconography and ideology, we examined them not in isolation but in terms of their relation to other aspects of culture, including sociopolitical organization, funerary ritual, public ceremonies, as well as militarism and imperial control—an approach that adheres to the multivariate forms of interpretation and explanation that Flannery has long advocated.

We do not argue for the primacy of either ideological or materialistic factors as the basis of power at Llano Perdido. The cacique's direct access to storehouses filled with wealth might seem, on the face of it, to reflect a straightforward material component to his power, but it is unlikely that such unequal control of wealth could have persisted for long here without ideologically sanctioned authority. We have proposed that an ideological complex involving ancestor veneration and the association with Lightning was critical to the legitimation of chiefly authority at Llano Perdido. To have any political effect, however, these ideas had to be transferred from the cognitive domain into the material world. The ideology was actualized through funerary ritual at a chief's interment and also through a pattern of regular public feasting and ceremony directed by the cacique, all employing vessels with the Lightning motif. These rituals presented the cacique as the focus of funerary ritual, the pivot of public ceremony, the intermediary between the villagers and the ancestral elite, and the sanctified representative of Lightning—a combination that surely made a formidable contribution to the legitimation of the cacique's authority.

We have also argued that the abrupt disappearance of the Lightning motif (and the funerary-feasting ceremonial complex associated with it) at La Coyotera between the Perdido and Lomas phases cannot be fully understood without taking into account the Zapotec takeover of the Cañada, which brought with it a new ideology of imperial domination associated with the jaguar and the *yàgabetoo*, along with a host of economic, social, political, and religious changes in the subjugated Cañada. In sum, throughout this chapter, we have endeavored to heed Flannery's wise counsel and use a holistic approach, one that "ties religion to social organization, politics, and subsistence, rather than leaving it on the ephemeral plane of mental activity" (Flannery, 1976:331).

REFERENCES

Burgoa, F. F. de, 1674a, *Geográfica descripción*, Vol. 1. Publicaciones del Archivo General de la Nación 25. Talleres Gráficos de la Nación, Mexico.

Burgoa, F. F. de, 1674b, *Geográfica descripción*, Vol. 2. Publicaciones del Archivo General de la Nación 26. Talleres Gráficos de la Nación, Mexico.

Caso, A., 1947, Calendario y escritura de las antiguas culturas de Monte Albán. In *Obras completas de Miguel Othón de Mendizábal*, Vol. 1, pp. 5–102. Mexico.

Caso, A., and Bernal, I., 1952, *Urnas de Oaxaca*. Memorias No. 2. Instituto Nacional de Antropología e Historia, Mexico.

Caso, A., Bernal, I., and Acosta, J., 1967, *La cerámica de Monte Albán*. Memorias No. 13. Instituto Nacional de Antropología e Historia, Mexico.

Castañeda, F. de, 1905, Relación de Teutitlán del Camino: Hecha el 15 de septiembre de 1581. In *Papeles de Nueva España, Segunda Serie, Geografía y Estadística, Tomo IV*, by F. Del Paso y Troncoso, pp. 213–231. Est. Tipográfico "Sucesores de Rivadeneyra," Madrid.

Demarest, A., 1989, Ideology and Evolutionism in American Archaeology: Looking beyond the Economic Base. In *Archaeological Thought in America*, edited by C. C. Lamberg-Karlovsky, pp. 89–102. Cambridge University Press, Cambridge.

Demarest, A., and Conrad, G. (editors), 1993, *Ideology and Pre-Columbian Civilizations*. School of American Research Press, Santa Fe.

DeMarrais, E., Castillo, L. J., and Earle, T., 1996, Ideology, Materialization, and Power Strategies. *Current Anthropology* 37:15–31.

Flannery, K. V., 1972, The Cultural Evolution of Civilizations. *Annual Review of Ecology and Systematics* 3:399–426.

Flannery, K. V., and Marcus, J., 1976, Formative Oaxaca and the Zapotec Cosmos. *American Scientist* 64:374–383.

Flannery, K. V., and Marcus, J., 1993, Cognitive Archaeology. *Cambridge Archaeological Journal* 3:260–270.

Flannery, K. V., Marcus, J., and Kowalewski, S., 1981, The Preceramic and Formative in the Valley of Oaxaca. In *Supplement to the Handbook of Middle American Indians, Volume 1: Archaeology,* edited by J. Sabloff, pp. 48–93. University of Texas Press, Austin.

Flannery, K. V. (editor), 1976, *The Early Mesoamerican Village*. Academic Press, New York.

Hodder, I., 1991, *Reading the Past*. 2d ed. Cambridge University Press, Cambridge.

Hunt, E., 1972, Irrigation and the Socio-Political Organization of the Cuicatec Cacicazgos. In *The Prehistory of the Tehuacán Valley, Volume 4: Chronology and Irrigation,* edited by F. Johnson, pp. 162–248. University of Texas Press, Austin.

MacNeish, R. S., Peterson, F., and Flannery, K. V., 1970, *The Prehistory of the Tehuacán Valley, Volume 3: Ceramics*. University of Texas Press, Austin.

Marcus, J., 1976, The Iconography of Militarism at Monte Albán and Neighboring Sites in the Valley of Oaxaca. In *The Origins of Religious Art and Iconography in Preclassic Mesoamerica,* edited by H. B. Nicholson, pp. 123–139. Latin American Center, University of California at Los Angeles, Los Angeles.

Marcus, J., 1978, Archaeology and Religion: A Comparison of the Zapotec and the Maya. *World Archaeology* 10:172–191.

Marcus, J., 1980, Zapotec Writing. *Scientific American* 242:50–64.

Marcus, J., 1983, Rethinking the Zapotec Urn. In *The Cloud People: Divergent Evolution of the Zapotec and Mixtec Civilizations,* edited by K. V. Flannery and J. Marcus, pp. 144–148. Academic Press, New York.

Marcus, J., 1989, Zapotec Chiefdoms and the Nature of Formative Religions. In *Regional Perspectives on the Olmec,* edited by R. J. Sharer and D. C. Grove, pp. 148–197. Cambridge University Press, Cambridge.

Marcus, J., and Flannery, K. V., 1978, Ethnoscience of the Sixteenth-Century Valley Zapotec. In *The Nature and Status of Ethnobotany,* edited by R. Ford, pp. 51–79. Anthropological Papers No. 67. Museum of Anthropology, University of Michigan, Ann Arbor.

Marcus, J., and Flannery, K. V., 1994, Ancient Zapotec Ritual and Religion: An Application of the Direct Historical Approach. In *The Ancient Mind: Elements of Cognitive Archaeology,* edited by C. Renfrew and E. Zubrow, pp. 55–74. Cambridge University Press, Cambridge.

Marcus, J., and Flannery, K. V., 1996, *Zapotec Civilization: How Urban Society Evolved in Mexico's Oaxaca Valley.* Thames and Hudson, London.

Marcus, J., Flannery, K. V., and Spores, R., 1983, The Cultural Legacy of the Oaxacan Preceramic. In *The Cloud People: Divergent Evolution of the Zapotec and Mixtec Civilizations,* edited by K. V. Flannery and J. Marcus, pp. 36–39. Academic Press, New York.

Miller, D., and Tilley, C. (editors), 1984, *Ideology, Power, and Prehistory.* Cambridge University Press, Cambridge.

Polaco, O., 1997, Appendix A. Shell Identifications. In *Archaeology of the Cañada de Cuicatlán, Oaxaca,* by C. Spencer and E. Redmond, pp. 611–613. Anthropological Papers No. 80. American Museum of Natural History, New York.

Pyne, N., 1976, The Fire-Serpent and Were-Jaguar in Formative Oaxaca: A Contingency Table Analysis. In *The Early Mesoamerican Village,* edited by K. V. Flannery, pp. 272–280. Academic Press, New York.

Redmond, E., 1983, *A Fuego y Sangre: Early Zapotec Imperialism in the Cuicatlán Cañada.* Memoirs No. 16. Museum of Anthropology, University of Michigan, Ann Arbor.

Redmond, E., and Spencer, C., 1982, Chiefdom and State in Formative Oaxaca. *Journal of the Steward Anthropological Society* 13:2:7–38.

Shanks, M., and Tilley, C., 1987, *Social Theory and Archaeology.* Polity Press, Cambridge.

Spencer, C., 1982, *The Cuicatlán Cañada and Monte Albán: A Study of Primary State Formation.* Academic Press, New York.

Spencer, C., and Redmond, E., 1997, *Archaeology of the Cañada de Cuicatlán, Oaxaca.* Anthropological Papers No. 80. American Museum of Natural History, New York.

Weitlaner, R., 1969, The Cuicatec. In *Handbook of Middle American Indians, Volume 7,* edited by R. Wauchope and E. Vogt, pp. 434–447, University of Texas Press, Austin.

Wilkinson, R. G., 1997, Appendix B. Human Skeletal Remains from La Coyotera. In *Archaeology of the Cañada de Cuicatlán, Oaxaca,* by C. Spencer and E. Redmond, pp. 614–620. Anthropological Papers No. 80. American Museum of Natural History, New York.

Chapter 7

Games, Players, Rules, and Circumstances

Looking for Understandings of Social Change at Different Levels

ROBERT D. DRENNAN

INTRODUCTION

Archaeologists study broader patterns of change in human societies over longer periods of time than do the practitioners of any other scholarly discipline. This is both the bane of our existence and the only real reason for it. Doing archaeology has many charms—almost all of them entirely unrelated to the objectives of archaeology—but finally it is an excruciatingly slow, painfully incomplete, and exorbitantly expensive way to find out about past human activities. It is, however, the only way to find out about many past human activities. Even for the quite recent past, the gaps and biases of historical records provide ample opportunity for archaeology to complement history. The longer the time span under study, the more critical archaeology is likely to be to the construction of an adequate account of what happened. Study of these very long sequences of social change looms large in archaeology. Such study is, after all, the one thing archaeologists can do better than anyone else. The longest sequences of human social change that are interesting to study include substantial periods for which historical records

ROBERT D. DRENNAN • Department of Anthropology, University of Pittsburgh, Pittsburgh, Pennsylvania 15260.

Cultural Evolution: Contemporary Viewpoints, edited by Gary M. Feinman and Linda Manzanilla. Kluwer Academic/Plenum Publishers, 2000.

are rudimentary or nonexistent and archaeology is the only source of information. The investigation, then, of the very broad patterns of change that occur over periods longer (often substantially longer) than the few centuries that can be studied reasonably adequately from historical sources of information is the unique contribution that archaeology has to offer.

The uniqueness of this contribution raises complicated issues for archaeology's relationship to other disciplines. These complications concern, among other things, the similarities between the processes of change that we study and those that are the focus of other social sciences (with their much shorter time perspectives) or that we casually can observe in our own daily lives. I take it for granted here that, as archaeologists, we are interested in improving our understandings of the processes of social change and of the forces that drive those processes. To the extent that we have some understanding of such processes, we should be able to say things, not only about what happened in the past, but also about how and/or why these things happened in the ways they did. Making a distinction between "how" and "why" is not important to the observations I make here. And I am intentionally avoiding the word *explain*, because I do not want to conjure up visions of science (with or without a capital *s*) at the expense of what many regard to be rather different ways of understanding processes.

The kinds of understandings of social change that I assume we are seeking, though, involve our ability to say things about how or why a particular society came to be the way it was (as opposed to some other way that it might have been but was not); to say things about how or why it is that some patterns of organization take rather similar forms repeatedly in a number of different and unrelated sequences; and to say things about how or why such similarities go only so far, giving way eventually to particular kinds of variation. At this point in the history of archaeology, I am unwilling to hazard a guess about what proportion of archaeologists would find this a congenial, although not necessarily complete, account of what it is we are ultimately after. This account is, in any event, my point of departure for what I intend as a series of observations about some recent theoretical trends, what they offer us, how they relate to older ideas, and what especially interesting issues remain unresolved. I do not pretend that these observations are comprehensive or even that they are entirely consistent with each other. I offer them as another load of bricks that may be useful as we continue our efforts to build a more adequate theoretical structure in which to understand the forces that drive social change and the ways in which these forces produce the results that they do.

ACTORS AND STRUCTURES

A great deal of the current (very healthy) ferment in archaeological theory can be attributed to a surge, perhaps a tsunami, of interest in individual actors in society and their impact on processes of change. (As a tiny example of this literature, one could cite Blanton et al., 1996; Brumfiel, 1992; Brumfiel and Fox, 1994; Cowgill, 1993; Earle, 1997; Hayden, 1995; Spencer, 1993.) Previously popular "cultural evolutionary" or "ecological" or "functional" theoretical approaches are commonly found wanting for focusing all their attention on social or cultural patterns and for treating individuals in society as automatons all behaving identically in a manner prescribed for them by their culture. (Marxist approaches comprise the other major theoretical family in archaeology—a broad family of such internal diversity that I do not attempt here to explore all its relations to the issues taken up.) Such critiques are clearly sometimes overdrawn, and examples to demonstrate this are easily found in what were at the time regarded as seminal works of the "new" cultural evolutionary-ecological-functional archaeology. Flannery (1968a), for instance, wove the role of individual decision making through the entire fabric of his influential account of agricultural origins in Tehuacán.

Cowgill (1975) criticized proponents of population pressure as a cause of social change for failing, among other things, to think of individual actors and their motivations. That is, population growth to a point of imbalance between population and basic resources on a regional (or even larger) scale is easy to think of as a macroproblem—a problem in the aggregate for the large number of mouths that need to be fed. Macroproblems, of course, have macrosolutions—in this case, perhaps, agricultural intensification that would increase the total regional agricultural production. One of Cowgill's (1975:506) complaints about this particular brand of thinking was that it ignored the fact that such changes in agricultural technology come about only if individuals (a lot of individuals) change their behavior. Cowgill judged the circumstances in which population pressure would place individuals to be very inimical to the kinds of changes in behavior that would be required to effect "agricultural intensification." He envisioned recurrent food shortages and intensified competition over the most productive and most reliable agricultural land—conditions that he thought likely to lead to hoarding and fighting than to collaboration in major capital projects whose ultimate effect would be to increase the total amount of food produced in the region.

In short, looked at the point of view of the individual enmeshed in it, the problem created by population pressure is not how to feed all the mouths in the region. Rather, it is how to feed oneself and one's family in increasingly difficult circumstances. The solution, from the individual's point of view, need not have anything at all to do with increasing regional agricultural production. For Cowgill, a much more likely solution from the individual's point of view was to take food, or the land on which to grow it, away from others. Cowgill, then, arrived at a fundamental contradiction between the glib macroanalysis and his microanalysis, that is, the ways in which it seemed reasonable to expect individual actors to act under certain circumstances.

There need not necessarily be such an inconsistency between macroanalyses and microanalyses, however, and population pressure can continue to serve as an interesting arena for thought experiments. Since people do have to eat, it should come as no surprise that producing the food required to feed a population has attracted theoretical attention. Population pressure has especially been advocated as a force pushing the hunter-gatherer populations of various regions toward agricultural subsistence, probably because the one advantage that agriculture seems to offer broadly over hunting and gathering in many parts of the world is the ability to increase food production in a given area. One can, however, quite reasonably inquire whether this "advantage" is one that would be perceived or appreciated by the inhabitants of any region and whether thinking in these terms at all is conducive to understanding how or why people modified their behavior in such a way as, collectively, to cease being hunter-gatherers and become agriculturalists.

It is not necessary to imagine that prehistoric hunter-gatherers recognized that regional population was growing, and that they could produce enough food to feed this larger number of mouths only if everyone started farming, in order to argue for population pressure as the causal force, or one among several causal forces, in the shift to agriculture. Indeed, no population pressure advocate has probably ever had such a mechanism in mind. Lack of attention to specific mechanisms at a microanalytic level, however, leaves the way open for us to impute such thinking to them. As applications of optimal foraging theory and related approaches to hunter-gatherer subsistence make clear, however, there are perfectly reasonable accounts of the ways in which regional population growth can change the calculus of subsistence decision making for the individual or the family (see, for example, Christenson, 1980; Earle, 1980).

For one thing, population growth will make some resources much more costly to exploit because they will become markedly scarcer. As population increases and hunting becomes more intensive, the populations of at least some species of animals will decrease, increasing the average time required to find and kill one of them and/or increasing the probability that a hunt for one of them will simply fail. Both of these changes will make hunting that species a less attractive option for hunter-gatherers, who, other things being equal, might be expected more often to choose other options that would, at lower regional population levels, be rejected. Similarly, as population increases, a group will more often be disappointed when arriving at a locale where abundant plant resources are expected because another group will have been there first and collected them. These kinds of hunting-and-gathering results will make exploitation of different species and different patterns of hunting and gathering, which might have been rejected earlier because they were more work, seem, by comparison, less unattractive than they had been. If these choices whose relative attractiveness increases are ones that lead in the direction of agriculture, then some causal role for regional population pressure can be seen to be more than just a macroanalytic just-so story with no connection to the forces that shape individual decision making.

I hasten to make clear that I use this example not to argue that population pressure was an important force in the origins of agriculture. Indeed I see the empirical case against this view as strong. The point is that there are at least some openings for dealing with population pressure successfully at a microanalytic level as well as at a macroanalytic level. We can address the question of why prehistoric hunter-gatherers, faced with decreases in the attractiveness of some of their major food sources, might choose, say, to intensify collecting of a plant that eventually became an important cultivar instead of making weapons and attacking the competing groups that were making traditional foods harder to get (either one of which options might be viewed as a plausible response at the microanalytic level to population pressure).

Although there are complicating factors, the issue of agricultural intensification can be viewed in a similar light, at least in instances where the nature of intensification does not absolutely require centralized control and/or management. In at least some instances, agricultural production can be intensified as a consequence of individual decisions to change cultivation techniques so that (usually with the application of more labor) more food is produced on a given plot of land. Regional population pressure might be felt by individuals primarily as a shrinking of the plots of land available to them for cultivation as landholdings are increasingly subdivided among more people. The effect would be to increase the costs and/or decrease the benefits of extensive cultivation, until its cost/benefit ratio was less favorable than that of cultivating more intensively. If a large number of individuals, perhaps quite independently, make a choice to cultivate more intensively, the result, viewed at the regional scale, is a macrosolution to a macroproblem. The numerous individual plots being cultivated in an intensive way are easy to talk about as a change in agricultural technology, and the net effect of this change in agricultural technology at the regional scale is the ability to feed a larger regional population.

The individual decisions made about how to feed individual families, then, simply add up to a solution (presumably for everyone) to the population-pressure problem. Once again, although making weapons and attacking competitors is another possible solution, we can discuss the circumstances in which this option might be less attractive to individuals than intensifying production on individual plots.

The reason that it is at least logically possible (although not necessarily empirically accurate) to view population pressure as a causal force in both macro- and microanalyses of changes in subsistence patterns is that both regional population and subsistence patterns can amount to nothing more than the sum of individual parts. That is, the regional population is simply the sum of all the individuals in the region, and the subsistence patterns or agricultural technology of a group is often simply the sum of a whole series of individual decisions. Neither necessarily has any more to it than that.

CHANGING THE RULES OF THE GAME

Many of the things whose emergence and development we would like to understand, however, we are accustomed to thinking of as having emergent properties. That is, they are cultural patterns that we ordinarily take to be more than simply the sum of their parts. For example, the shift from what we have traditionally called egalitarian social organization to hereditary social ranking is not easily conceived as coming about as individuals at different rates and in different ways, for whatever their reasons, gradually come to defer to the members of a particular family more and more often and to treat them as equals less and less often. Hereditary social ranking is not so much a matter of individual decisions about behavior as it is a collective pattern from the start. We are at least accustomed to seeing it as a change in structure, rather than as a shift in patterns of individual actions.

The emergence of hereditary social ranking, moreover, requires a change in norms about how individuals decide what to do—a change from a basic rule of *not* deferring to people just because of the family they were born into to a basic rule of deferring to people *just* because of the family they were born into. In the previous example of changing subsistence patterns as a response to population pressure, the norms according to which people make their subsistence choices do not change: it is supposed throughout that they choose to pursue subsistence options that offer more favorable cost/benefit ratios. (Once again, this latter may or may not be empirically accurate; the point here is that, if empirically accurate, the account offers not only a description but also some degree of understanding of the processes involved.)

Another way to frame this is that a shift from hunting and gathering to agriculture or an intensification of agriculture as the sum of a series of basically independent decisions by individual actors works quite simply and directly because individual actors can be seen to be playing the same game by the same rules before, during, and after the change. That is, individuals are taken to make their decisions on the basis of self-interest, evaluated in terms of the cost/benefit ratios of feeding their families. Population growth brings about changes that affect the relative cost/benefit ratios of different options available to them, so they decide to change their tactics, but the rules of the game they are playing have not changed.

We are, however, accustomed to view changes such as the emergence of hereditary social hierarchy as a fundamental change in the rules by which the social game is played. Sahlins (1963), in his classic discussion of the divide between big-man and chiefly organization, was much concerned with precisely this issue: how much the rules of the social game must change if the one is to change into the other. He saw the rules of the game in big-man society as fundamentally incompatible with the emergence of chiefs. The very patterns of behavior that enabled a person to become a big-man prevented an ambitious big-man from converting himself into a chief. For Sahlins, generosity and giving were the core of successful competition to become a big-man, while hoarding and accumulation were essential to chiefly competition. The minute a big-man was seen to be accumulating wealth for himself, he lost respect and his faction of followers. The big-man who wished to make a leap to an even more exalted position, then, was faced with a serious structural obstacle. The change could not come about as a consequence of changing circumstances that altered individuals' independent decisions about how to behave—decisions made on the basis of the same rules as before. It had to involve a change in the norms of individual behavior.

Brumfiel (1994:4–5), in focusing strongly on the role of individual actors' decision making in the context of the formation of and competition between factions, has also adopted this perspective. She takes the kind of long-term social change that archaeologists focus on to comprise "social transformation"—change in the rules by which factions compete and by which individuals decide which faction to ally themselves with. In the process of this change, new opportunities are created for factions competing with other factions, for leaders competing for ascendance, and for individual followers affiliating themselves with factions. To put it in Sahlins's terms for one example, although Brumfiel does not, there comes a time when an ambitious big-man does not lose out by breaking the old rules, but instead wins, and wins big. For some reason, breaking the norms of behavior does not cause him to lose his faction but enables him to cement his faction even more closely to himself, enlarge it, defeat rival factions, and make his ascendance considerably more permanent for himself and his heirs. Spencer (1993, 1994) thinks of this same social transformation in precisely these terms and hypothesizes about the circumstances that make it possible.

To conceive of major changes in this way, however, introduces a serious complication into the logic of analyzing the processes behind them simultaneously and harmoniously at both micro- and macroscales. Instead of seeing how changes, easy to conceptualize at the macroanalytic scale, alter the circumstances in which individuals, playing by the same rules throughout, change their behavior, we now must ask how it is that the rules according to which individuals make their decisions get changed. There would seem to be two approaches to answering such a question.

One approach, raised obliquely by Brumfiel (1992:558–560, 1994:12) in advocating an "alternat[ion] between a subject-centered [i.e., micro] and a system-centered [i.e., macro] analysis," finally depends on especially skillful aspiring leaders who occasionally are able to succeed at precedent-breaking strategies that alter the circumstances in which subsequent competitions are played out. Steponaitis (1991:227) arrives at a similar point, although with considerably more regret than Brumfiel displays. If the course of long-term social change does depend heavily on the stepwise accumulation of the consequences of that particular leaders' successes at cajoling or bamboozling their followers into changing the rules according to which they behave, then the courses of particular sequences of change are idiosyncratic, or, in Brumfiel's (1994:12) words, "uniquely complex and contingent." There would seem to be little we can say as a matter of general principle about how they vary. Along this route, a focus on individual actors and competition between the factions into which they choose to form themselves might well be empirically accurate, but it offers little comprehension of processes. Indeed, it causes us to set our sights considerably lower than a generation of processual archaeologists have done. In fact, this approach, rather than enhancing our comprehension of processes of social change, is largely equivalent to the great-man history of an even earlier generation.

THE CIRCUMSTANCES IN WHICH THE GAME IS PLAYED

There is a second (and contrasting) approach to understanding the critical events in which aspiring leaders manage to win out over competitors in new and different ways by breaking the rules and thereby institute new rules. We can focus on the circumstances that enable a leader to accomplish such a thing. This approach does not depend on the occasional flash of leadership genius. Instead, a steady supply of eager competitors for power and prestige is assumed. They are just always there in human society, winning and losing and trying continually to stretch the envelope. Only under certain circumstances, however, can even a very able, aspiring leader win by breaking the old rules and thus institute new ones.

In one example of such an approach, Spencer (1993, 1994) has argued that, when circumstances permit a big-man-type leader to exert control over human affairs in dimensions both internal and external to his region, he is in a position to convert himself into a chief whose hierarchical position can be passed on to his descendants. As long as the leader's impact is felt only in the internal arena (as, for example, in organizing agricultural production) or in the external arena (as, for example, in acquiring goods from outside the region or leading his group in warfare against neighboring groups), then the big-man cannot make the jump to chief. That is, he cannot induce the members of his faction to treat him as a chief, and, more important, his heir cannot induce the members of the faction to treat him as a chief just because of his descent from the previous leader. For Spencer, the critical circumstance that enables a leader to operate forcefully in both internal and external realms is risk—risk that he attributes to environmental fluctuations and their impact on agricultural production, or to warfare, or potentially to other factors. This view offers comprehension of processes by specifying something about when and where one particularly important feature of hereditary social hierarchy can be expected to appear: where some element of risk offers aspiring leaders a very special opportunity.

Clark and Blake (1994) have offered a structurally similar explanation of essentially the same transition in social hierarchy. For them, however, it is not risk that is the crucial element, but rather the opposite situation: especially rich, highly dependable resources. These are the circumstances they see as offering aspiring leaders the opportunity to accumulate resources they can use to build their factions and their own powers to previously unheard of heights. They can, moreover, pass these resources on to their heirs, who can use them in the same way and thus inherit high position in the social hierarchy.

Both Spencer, on the one hand, and Clark and Blake, on the other, offer their formulations as approaches to the emergence of hereditary social hierarchy that refocus attention on the actions of individuals in a productive way and provide both empirically as approaches to the more accurate and theoretically more satisfying understandings of the dynamic of this important social transformation. Spencer is more explicit than Clark and Blake in seeking such understandings, not by turning attention away from the structural considerations and toward individual actors, but by seeking to add "human agency" to models that also include attention to the way in which circumstances constrain the possibilities for individual decisions.

Models like these two offer us understanding of the dynamic of social change in several different ways. One kind of understanding has directly to do with the forces that drive change. The existence of an urge to compete for resources, power, prestige, or some other manifestation of exalted position, in at least some portion of any human population, is evidently assumed in such models, although this assumption is not always made explicit. Were it not for such an urge, there would be no motor to drive change in either Spencer's or Clark and Blake's model. The emergence of hierarchy is due to the desire to dominate in some way, but a self-aggrandizer cannot dominate without some structure of unequal relations. It is the process of successful self-aggrandizement, then, that creates the unequal social, political, economic, or other relationships that form a hierarchy. This view, and the notion that the urge to self-aggrandize is just part of human nature, are consistent with recent arguments that varied kinds of unequal relationships are present in small-scale hunter-gatherer societies traditionally branded as egalitarian in the cultural evolutionary literature (see, for example, Flanagan, 1989; Gregg, 1991; Price and Brown, 1985). The main force that produces social hierarchy, then, seems to operate anywhere and everywhere, and all societies are seen, at least metaphorically, as striving toward the kinds of inequalities that have proven such a pervasive organizational device in large-scale societies.

We are provided, then, with a motor, always left running, that will drive human societies toward hierarchical organization, and this would certainly seem to be an important element in any understanding of the emergence and development of hierarchies. On the other hand, thus far in such an account, we have not addressed the issues of understanding how it is that some societies do not develop hierarchical organization (or if we see at least some kind of hierarchy everywhere, how it is that some societies develop only the most modest of inequalities) while in others dramatic disparities of prestige, power, and/or wealth emerge. If the force is a feature of human nature—in effect, a constant—it is not possible to understand variation by reference to it alone. This view is perhaps consistent with the perspective, offered by Marcus and Flannery (1996:244–245), that a focus on human actors is most useful in considering single sequences of social change but that other kinds of approaches are more useful in comparative studies. This may be another way of framing the reasons why Spencer (1993) finds it important to try to combine such a "human agency" approach with a consideration of the impact on the outcomes that the circumstances in which individuals act may have.

Spencer's focus on the importance of risk to the development of hereditary social hierarchy is an appeal to the circumstances in which self-aggrandizers act in an effort to understand variation—how it is that hereditary social hierarchy emerges when and where it does and not in other places and times. Agricultural risk creates opportunities for leaders to do things to mitigate the risk (construct and manage irrigation systems or drained field systems, for example) and thus to benefit their followers. Providing such a benefit is, of course, a powerful attraction to followers, and we can easily envision a dynamic of individual decision making from the perspective of both leaders and followers. Leaders seek to expand their factions and exert stronger control over their followers; followers seek to join factions where the cost of membership (in terms of ceding control to leaders) is relatively low compared with the benefits of membership (for example, in terms of enhanced security from food shortage). A leader who can provide a more valuable benefit can also exact higher costs from his followers. The element of risk, then, is what permits leaders to cross some threshold in the strength and permanence of their leadership. Agricultural risk is only one among a number of kinds of risk that might have this same general impact on the course of social change. Another that Spencer also develops is risk from warfare.

Clark and Blake's focus on the importance of lush and predictable resources is a similar appeal to the circumstances in which the urge to dominate plays itself out. The same urge to self-aggrandize is seen to be limited if resources are sparse or subject to the kinds of risks that for Spencer create the circumstances in which leaders can push social hierarchy to new heights. In contrast, if resources are abundant, predictable, and easily exploited, providing taxes or tribute to leaders imposes relatively light burdens on followers, and it should be easier for leaders to extract such surpluses from their followers.

Clark and Blake, on the one hand, and Spencer, on the other, thus present alternative and apparently conflicting models about the circumstances that most encourage the development of hereditary social hierarchy. Each model offers the same kind of understanding of the process. Both rely on the individual initiative of aspiring leaders to dominate others and build factions as the driving force behind this social change. Both are readily comprehensible in the microanalytic terms of the motivations and individual decision making of both leaders and followers. Everyone (or at least some substantial proportion of the population) wants to dominate others in some way or other, but everyone is willing to kowtow to some other self-aggrandizer when losing out in the pecking order and as long as the costs of kowtowing seem modest compared with the benefits provided. Spencer's suggestion of ideal circumstances works because the risk-reduction benefits provided by a leader's management of social affairs are especially high. The circumstances that Clark and Blake focus on are ideal because the costs of providing the "surpluses" that leaders demand are relatively low.

EMPIRICAL EVALUATION AND THE CONTRIBUTION OF ACTION THEORY

This contrast between the models suggested by Spencer and by Clark and Blake offers an ideal opportunity for empirical evaluation. If Spencer is right, then hereditary social hierarchy should generally develop in circumstances of substantial risk. In contrast, if Clark and Blake are right, then hereditary social hierarchy should generally develop in circumstances of abundant, predictable, easily exploitable resources—that is, in circumstances of very little risk. Spencer, indeed, follows just this route in the two empirical examples he provides of his model in operation—both examples of situations in which he finds evidence of considerable risk, evidence of the development of hereditary social hierarchy, and evidence of emerging leaders engaging in the kinds of organizational and management activities that would reduce their followers' risk level. The other side of this empirical coin would be to see whether in areas of substantially lower risks the development of hereditary social hierarchy does not occur (or at least proceeds slowly and/or to a limited extent). Spencer does not attempt this part of the empirical evaluation of his model, but the empirical case discussed by Clark and Blake could be cast in such a role. They see early and rapid development of hereditary social hierarchy in a region that to them has the opposite of the characteristics Spencer would like to see. This case could count as empirical evidence against Spencer's model, and in precisely the same way Spencer's two cases could count as empirical evidence against Clark and Blake's model.

At this point, we should ask how new these approaches are and just what contribution an emphasis on individual actors makes to them. The notion that high-risk environments are conducive to the development of complex patterns of organization that generally involve social hierarchy is not entirely novel (Braun and Plog, 1982; Brose, 1979; Halstead and O'Shea, 1982; Isbell, 1978; Kelly, 1991; Peebles and Kus, 1977; Steponaitis, 1983; and many others). The notion that lush environments are conducive to the development of social hierarchy has perhaps not been so prominent, although Anderson, Stahle, and Cleaveland (1995) as well as Sebastian (1991) rely on it, and Sanders and Webster (1978) played these two apparently opposite environmental characteristics off in an interesting way. They argue that easily exploited zones with little environmental risk lead to rapid initial development of status hierarchies that then collapse or stagnate, while high-risk zones, where organizational solutions to reducing risk and increasing production are appropriate, lead eventually to much greater elaboration of hierarchical organization. Sanders and Webster's approach has been argued to be empirically inaccurate (Drennan, 1991), but it does suggest a way to go beyond the relatively simplistic terms of models that offer a definition of Conditions *X* in which hereditary social hierarchy develops, and by implication, at least, of Conditions *Y* in which it does not. A model that attempts to tell us how it is that one sequence of developing social hierarchy takes one path and another sequence follows a different trajectory offers a richer and fuller understanding.

Sanders and Webster, however, in presenting their model, do not engage in a lengthy discussion of motivations and decision making at the individual level. In this regard, they do not, perhaps, differ from the long tradition of cultural evolutionary work. This may be attributable to a focus on understanding similarities and differences, which makes attention to constants (like an urge to dominate that is just part of human nature) superfluous. But for accounts that focus only on the circumstances in which things happen to make any sense at all, they must be provided, at least implicitly, with some notion of the mechanisms at work. Clark and Blake's or Spencer's focus on individual actors makes explicit one way in which models focused on circumstances may work by providing what amounts to an account of proximate causes. At the level of empirical evaluation, though, the approach discussed above does not seem very different from the traditional cultural evolutionary literature.

Moreover, there is a degree of reliance on environmental variables that has led to the branding of such approaches as environmentally deterministic (e.g., Feinman, 1995). Spencer, perhaps, more than Sanders and Webster or Clark and Blake, among the examples discussed here, partially escapes from this charge by focusing not exclusively on environmental risk but also on risk from such factors as warfare. Unless we attribute warfare to demographic pressure, we are left with the question (raised but not answered by Brumfiel, 1994:7) of why warfare occurs in some societies but not others. It is disconcerting to arrive finally at such reliance on environmental factors in individual-competition models because the literature focusing on individual decision making tends to see itself as offering escape from the inadequacies of an outmoded environmentally deterministic approach.

The tack taken by Spencer in evaluating his model, and on which we have just been reflecting, differs from the empirical evaluation suggested by Clark and Blake (1994:22–30) for their model. Although it seems to be logically implied by their argument, Clark and Blake do not focus on comparative analysis of environmental richness and predictability, investigating whether it is generally the case that hereditary social ranking emerges under such conditions. Instead, they dedicate most of their attention to arguing that the particular nature and timing of changes in ceramic technology that accompanied their test case are attributable to competition for prestige and faction building by aspiring elites by way of public feasting. They attempt a different kind of empirical evaluation altogether. The question they attempt to answer empirically is not about the circumstances in which hereditary social hierarchy appears at all. Instead their question directly concerns the force driving the development of social hierarchy. Clark and Blake's empirical evaluation is less clear-cut than Spencer's. This is not to say that Clark and Blake's test necessarily arrives at erroneous conclusions, but simply that it is not fully developed. The empirical question they ask lacks clearly identifiable alternative answers.

Spencer's empirical evaluation is driven by the question, "Does hereditary social hierarchy develop in conditions of high risk or low risk?" The possible contradictory answers to this question are clear and unequivocal (although we could, of course, argue for a long time about the assignment of regions to these categories—that is a different issue). Considering Spencer's argument together with Clark and Blake's helps clarify the alternative answer, because they offer a specific and positive account of a kind of low-risk environment that would spur the development of social hierarchy. Clark and Blake's own empirical evaluation, however, is driven by the question, "Is the force behind the emergence of social hierarchy individual self-aggrandizement and competition?" The possible answers to this (intentionally oversimplified) question, of course, are "yes" and "no," but it is not so clear just what alternative force "no" might indicate. Here, contrasting with Spencer's model is no help, because the driving force in Spencer's model is the same as the driving force in Clark and Blake's.

Considering Spencer's and Clark and Blake's suggestions makes it clear that a focus on individual actors is not necessarily an alternative to a consideration of the circumstances in which individuals act. This observation is, of course, entirely consistent with the perspective adopted by Spencer, at least, who takes considerable and explicit pains to put the two perspectives together seamlessly. But just what contribution does a focus on individual actors make? How does using the urge to self-aggrandize as the driving force behind this social change represent an improvement over the other options? Indeed, what are the other options?

Perhaps the likeliest is a sort of group-selection argument based on the benefits of leaders' organizational or management efforts. In such an account, groups' variable success in their efforts to feed and reproduce themselves would lead to the expansion of some groups at the expense of others, and those that were best organized to cope with whatever were the critical circumstances would finally win out. After a couple of decades of vigorous attention to the weaknesses in group-selection arguments, there seems to be little support for this account and little reason to take it seriously as an alternative (although Dunnell [1980], for example, has argued that certain features of the development of complex society make group-selection arguments especially applicable here).

CONCLUDING OBSERVATIONS

At this point, we might well wonder whether attention to individual actors and their motivations offers as entirely new an angle on the processes of social change as some of its proponents seem to think or whether its contribution is much more modest—an account of proximate causes basically consistent with much of the macroanalysis of the cultural evolutionary literature of the past 40 years (for this really seems to be the perspective that Spencer offers us). I would like to close with a series of six interrelated (although perhaps not entirely uncontradictory) observations relating to this quandary.

First, construing human beings as innately endowed with an urge to acquire power, prestige, and/or wealth does indeed provide a coherent account of proximate causes. Most of the cultural evolutionary-ecological-functionalist literature of the 1950s through the 1970s can accurately be criticized for paying little explicit attention to analysis at this level. The classic processual models of archaeology, focusing on the particular factor or combination of factors that, for example, brought about social stratification, were invitations to understanding, but only on the basis of "bring your own proximate causes." The propitious circumstances they described for the emergence of social stratification could affect such change only if they provided special opportunities for self-aggrandizers to succeed or if they provided for the survival of hierarchically organized societies at the expense of others, and these were the two sets of mechanisms readers generally either provided or implicitly accepted. Action theory makes choice of the former explicit, and explicit exploration of this dynamic certainly enriches the accounts we can provide of long-term social change.

This leads to the second observation: one of the ways in which explicit attention to individual decision making enriches our accounts is by offering at least some possibility of a harmonious union between macro- and microanalytic levels. Although Spencer (1993, 1994) does not frame it in these words, his risk model is easy to conceptualize at the macroanalytic level. A regional population's successful solution to its large-scale problem is to increase agricultural production and make it more reliable through agricultural intensification. This approach integrates seamlessly into a microanalysis because a leader who can provide this kind of managerial solution to individuals' subsistence insecurity will be very successful at building a faction and will win out over competitors. This success comes about as individuals make decisions about which factions to affiliate with, accepting the costs and benefits associated with such affiliation (or lack of affiliation) after deciding that the cost/benefit ratio is the best to be obtained.

The third observation is a direct implication of the second: this union of macro- and microanalytic levels works only because we have, without realizing it, gotten beyond the problem of transformation or changing the rules of the game. In the terms just set forth, the rules of the game have not changed. Self-aggrandizers continue to strive for power, prestige, and/or wealth, using whatever tactics seem to work, playing the same game by the same rules as before. Followers continue to choose which factions to join based on cost/benefit ratios, playing the same game by the same rules as always before. At this level and by these rules, the transformation from big-men, who gain prestige by their generosity, to chiefs, who accumulate personal wealth is not seen as a change in the rules of the game. Individual players affiliate with the factions of those that provide the biggest benefits for the smallest costs, and the metaphor is one of a playing field where a much broader and more general set of rules subsumes the specific differences between big man and chiefly organization. It is important to note the fundamental importance to this perspective of the empirical observation that the seeds of social hierarchy are present everywhere in at least small ways. The questions, in all human societies, must be to whom to kowtow and how much to kowtow but never whether to kowtow at all, or this approach does not work. Whatever advantage this much broader view carries with it, however, it is not without its disadvantages.

And this brings us to the fourth observation: focusing on individual actors playing an almost cosmic game of one-upmanship or deciding whose star to hitch their wagons to is not very helpful for understanding variation. Since the urge to be kowtowed to (and the corresponding willingness under some circumstances to kowtow) are taken to be universal features of human nature, it is logically impossible to understand much about variation in the trajectories of long-term social change by reference to these features alone—they do not vary. At the same time, paradoxically, the ways in which individuals seek to pursue the urge to be kowtowed to provide the ultimate source of all the variability in the sequences we study. Models focused on human agency, however, do not offer us understanding of the generation of this variability but rather of its expression in changing social patterns. The understanding that is offered comes, not from the focus on human agency, but from attention to the circumstances in which those individual actors act.

This is not the same as saying that competition between aspiring leaders and their factions provides the fulcrum for "understanding the course of local history" (Brumfiel, 1994:12) but "is less useful for comparing" (Marcus and Flannery, 1996:245). Let's explore the game metaphor a little further. If we think of long-term social change as a giant game of chess, the driving force is the desire to win by taking the opponent's king. It is clearly necessary to take cognizance of this objective even to give a coherent description of a single game. But taking it fully into account does not help us "understand" one bit better how or why that game played itself out in the way that it did—as opposed to the myriad other ways it might have played itself out. Different chess games on the same board between the same two opponents take notoriously different courses with different outcomes. Each is probably, in Brumfiel's (1994:12) words, "uniquely complex and contingent." This variability is perhaps not the kind we should even seek to "understand," however, although it is certainly rewarding and worthwhile to describe it coherently.

It is a different kind of variability in social trajectories that we might more productively seek to understand (and of which the models I have focused on here offer us understanding). We can make the chess game metaphor do for this other level of understanding only if we extend it, by imagining a kind of hyperchess in which the size and shape of the board and the numbers of pieces of different kinds given each player can vary. This hyperchess would, of course, add another dimension of variability to a series of chess games—a dimension more amenable to "understanding." It is no more likely that any two chess games would be alike, but one might expect certain commonalities between the games played out on boards of the same shape and size and with the same set of pieces in play. We could "understand" variation at this level with reference to the variation in configuration of the board and the pieces. In short, some of this variation might be comprehensible with reference to the circumstances (i.e., the board and pieces) under which the game was played. It is in precisely this way that Brumfiel (1992, 1994), Spencer (1993, 1994), and Clark and Blake (1994) offered us understanding of long-term social change: by hypothesizing about the (alas, largely environmental) circumstances under which the social game can take a course of particular interest to the analyst. (Although the variation to be understood here is largely conceived of as a critical transformation that either does or does not take place rather than more subtle qualitative variation in trajectories [cf. Drennan, 1991, 1996].)

But how novel an understanding is it? "Complex political institutions might be precluded by ecological conditions that do not meet their own institutional requirements" (Brumfiel, 1994:6) is, finally, not such a different pronouncement from the generalization that tropical lowland environments cannot sustain complex societies (Meggers, 1954). Is it a major advance to ponder whether social stratification develops most readily in rich predictable environments (à la Clark and Blake [1994]) or in highly risky ones (à la Spencer [1993, 1994])? It sounds uncomfortably close to a much older contrast between, on the one hand, the hoary notion that the development of a stable, dependable agricultural subsistence made it possible for "man" to "advance" (e.g., Childe, 1951), among the achievements of which was included social hierarchy, and, on the other hand, the idea that the trappings of social hierarchy served as a means of banking agricultural surpluses as protection against future risks (Flannery, 1968b:107–108; Halstead and O'Shea, 1982).

The older incarnations of these ideas were developed without benefit of much attention to the individual players in the games they described. Thinking of such variability in human agency terms leads Brumfiel (1996:48) to ask, "Why do leaders initially opt for one form of strategy over another? Is this choice due to differences in available resources, demographic patterns, existing social structure, or value systems and cosmologies?" The existing situation, of course, does condition the possibilities for future development, but finding an understanding of variation in the social or cultural nature of an earlier period (the "existing social structure, or value systems and cosmologies") only leaves us with the question of how the system got to that point. One must imagine pursuing understanding farther and farther back into earlier periods until the variation we seek to comprehend does not depend primarily on still earlier social or cultural variation. On Brumfiel's list, only "available resources" and possibly "demographic patterns" allow escape from the infinite regression of accounting for cultural variability by reference to previous cultural variability, and these pull us back again toward an environmental determinism we sought to overcome by heading down this path in the first place.

The fifth observation is that this very broad perspective on individual players playing a game with the aim of advancing their own self-interest in whatever way works can help us overcome an old dichotomy of "managerial" versus "control" (Earle, 1987) or "voluntaristic" versus "coercive" (Carneiro, 1970) theories about the development of complex patterns of sociopolitical organization. Aspiring leaders build factions and advance their own interests in controlling their followers, at least in part, by providing managerial benefits to the members of their factions. Failure to attend carefully to the cost/benefit ratio for followers spells doom for a self-aggrandizer, although under some environmental or other circumstances the costs of leaving the faction can be quite high, making it possible for leaders to extract high costs for membership. The observation that self-aggrandizers serve their own interests by providing benefits (even if small ones) to those whose "surplus" they "expropriate," of course, is also the underlying rationale of "trickle-down economics" (characterized as "voodoo economics" by George Bush before he, himself, recalculated the cost/benefit ratio of membership in the Ronald Reagan faction he had previously opposed).

This approach casts all human social affairs into an economic mode of analysis, although perhaps not in all the ways sometimes thought. It does not, for example, impute explicit cost/benefit analysis to all the actors. They do, however, make all their decisions in ways that are functionally equivalent to cost/benefit decisions, just as I do when I decide not to go out to a movie because I have a lecture to write. I could certainly decide to go out, but the cost of either going to class unprepared or staying up all night to write the lecture outweighs the benefits of going to the movie. Neither costs nor benefits here are readily statable in economic terms, but the analysis is still, in effect, a cost/benefit one. Cowgill (1993) recognized that such cost/benefit analyses can include the evaluation of all manner of things that go beyond the strictly economic, but that, even allowing for this, there are severe limitations to treating all individuals as rational decision makers according to any prescribed list of criteria.

The sixth, and final, observation is not original. Modeling the dynamic of long-term social change as nothing more than the outcome of individual decision making in the framework of playing an economic game based on cost/benefit analysis "might be criticized as a projection of the competitive, self-seeking, pragmatic ideology of Western capitalist society" (Brumfiel, 1994:12). Brumfiel then goes on to suggest that "the problem is ameliorated by situating self-interested competition in specific ecological, social, cultural, and historical contexts." The principal payoff from this family of models appears to come, however, from construing all possible individual decisions as functional equivalents of cost/benefit decisions in a game of self-interest. Whether this turns out to be astute analysis or ethnocentrism, it is empirically alarming that a surge in the popularity of such an approach comes precisely at a time of unprecedented global ascendance of Western capitalism—from former socialist systems in disarray, to previously unthinkable levels of privatization of former public services on all continents, to the increasingly common internal and external pressures for institutions like universities to organize and direct themselves according to business principles.

Let me hasten to add that I do *not* conclude on this note to make a point about political correctness. I carry no brief for evaluating theoretical approaches in archaeology according to the portion of the political spectrum they are perceived to come from (cf. Flannery and Marcus, 1994). But as intellectual self-aggrandizers competing in our own game of power and prestige, are we nothing more than so many academic corks bobbing along on powerful cultural currents that condition (much more strongly than we are willing to believe) the rules that determine who wins and who loses? It startles me to think that even the current popularity of action theory in archaeology could be, in some poorly understood way, epiphenomenal to much broader cultural shifts. If this were so, then we would have reason to worry that we are overreaching in attributing to human agency the central dynamic role in the much more sweeping long-term social reorganizations whose study is the special province of archaeology.

REFERENCES

Anderson, D. G., Stahle, D. W., and Cleaveland, M. K., 1995, Paleoclimate and the Potential Food Reserves of Mississippian Societies: A Case Study from the Savannah River Valley. *American Antiquity* 60:258–286.

Blanton, R. E., Feinman, G. M., Kowalewski, S. A., and Peregrine, P. 1996, A Dual-Processual Theory for the Evolution of Mesoamerican Civilization. *Current Anthropology* 37:1–14.

Braun, D. P., and Plog, S., 1982, Evolution of "Tribal" Social Networks: Theory and Prehistoric North American Evidence. *American Antiquity* 47:504–525.

Brose, D. S., 1979, A Speculative Model of the Role of Exchange in the Prehistory of the Eastern Woodlands. In *Hopewell Archaeology: The Chillicothe Conference*, edited by D. S. Brose and N. Greber, pp. 3–8. Kent State University Press, Kent, OH.

Brumfiel, E. M., 1992, Distinguished Lecture in Archeology: Breaking and Entering the Ecosystem—Gender, Class, and Faction Steal the Show. *American Anthropologist* 94:551–567.

Brumfiel, E. M., 1994, Factional Competition and Political Development in the New World: An Introduction. In *Factional Competition and Political Development in the New World*, edited by E. M. Brumfiel and J. W. Fox, pp. 3–13. Cambridge University Press, Cambridge.

Brumfiel, E. M., 1996, Comment. *Current Anthropology* 37:48–50.

Brumfiel, E. M., and Fox, J. W. (editors), 1994, *Factional Competition and Political Development in the New World*. Cambridge University Press, Cambridge.

Carneiro, R. L., 1970, A Theory of the Origin of the State. *Science* 169:733–738.

Childe, V. G., 1951, *Man Makes Himself*. Watts, London.

Christenson, A. L., 1980, Change in the Human Niche in Response to Population Growth. In *Modeling Change in Prehistoric Subsistence Economies*, edited by T. K. Earle and A. L. Christenson, pp. 31–78. Academic Press, New York.

Clark, J. E., and Blake, M., 1994, The Power of Prestige: Competitive Generosity and the Emergence of Rank Societies in Lowland Mesoamerica. In *Factional Competition and Political Development in the New World*, edited by E. M. Brumfiel and J. W. Fox, pp. 17–30. Cambridge University Press, Cambridge.

Cowgill, G. L., 1975, On Causes and Consequences of Ancient and Modern Population Changes. *American Anthropologist* 77:505–525.

Cowgill, G. L., 1993, Distinguished Lecture in Archeology: Beyond Criticizing the New Archeology. *American Anthropologist* 95:551–573.

Drennan, R. D., 1991, Pre-Hispanic Chiefdom Trajectories in Mesoamerica, Central America, and Northern South America. In *Chiefdoms: Power, Economy, and Ideology*, edited by T. Earle, pp. 263–287. Cambridge University Press, Cambridge.

Drennan, R. D., 1996, One for All and All for One: Accounting for Variability without Losing Sight of Regularities in the Development of Complex Societies. In *Emergent Complexity: The Evolution of Intermediate Societies*, edited by J. E. Arnold, pp. 25–34. Archaeological Series No. 9. International Monographs in Prehistory, Ann Arbor.

Dunnell, R. C., 1980, Evolutionary Theory and Archaeology. In *Advances in Archaeological Method and Theory*, Vol. 3, edited by M. B. Schiffer, pp. 35–99. Academic Press, New York.

Earle, T. K., 1980, A Model of Subsistence Change. In *Modeling Change in Prehistoric Subsistence Economies*, edited by T. K. Earle and A. L. Christenson, pp. 1–29. Academic Press, New York.

Earle, T. K., 1987, Chiefdoms in Archaeological and Ethnohistorical Perspective. *Annual Review of Anthropology* 16:279–308.

Earle, T. K., 1997, *How Chiefs Come to Power: The Political Economy in Prehistory*. Stanford University Press, Stanford, CA.

Feinman, G. M., 1995, The Emergence of Inequality: A Focus on Strategies and Processes. In *Foundations of Social Inequality*, edited by T. D. Price and G. M. Feinman, pp. 255–279. Plenum Press, New York.

Flanagan, J. G., 1989, Hierarchy in Simple "Egalitarian" Societies. *Annual Review of Anthropology* 18:245–266.

Flannery, K. V., 1968a, Archeological Systems Theory and Early Mesoamerica. In *Anthropological Archeology in the Americas*, edited by B. J. Meggers, pp. 67–87. Anthropological Society of Washington, Washington, D.C.

Flannery, K. V., 1968b, The Olmec and the Valley of Oaxaca: A Model for Inter-Regional Interaction in Formative Times. In *Dumbarton Oaks Conference on the Olmec*, edited by E. P. Benson, pp. 79–110. Dumbarton Oaks, Washington, D.C.

Flannery, K. V. and Marcus, J., 1994, On the Perils of "Politically Correct" Archaeology. *Current Anthropology* 35:441–445.

Gregg, S. A. (editor), 1991, *Between Bands and States*. Occasional Paper No. 9. Center for Archaeological Investigations, Southern Illinois University, Carbondale.

Halstead, P., and O'Shea, J., 1982, A Friend in Need Is a Friend Indeed: Social Storage and the Origins of Social Ranking. In *Ranking, Resource, and Exchange: Aspects of the Archaeology of Early European Society*, edited by C. Renfrew and S. Shennan, pp. 92–99. Cambridge University Press, Cambridge.

Hayden, B., 1995, Pathways to Power: Principles for Creating Socioeconomic Inequalities. In *Foundations of Social Inequality*, edited by T. D. Price and G. M. Feinman, pp. 15–86. Plenum, New York.

Isbell, W. H., 1978, Environmental Perturbations and the Origin of the Andean State. In *Social Archeology: Beyond Subsistence and Dating*, edited by C. L. Redman, M. J. Berman, E. V. Curtin, W. T. Langhorne Jr., N. M. Versaggi, and J. C. Wanser, pp. 303–313. Academic Press, New York.

Kelly, R. L., 1991, Sedentism, Sociopolitical Inequality, and Resource Fluctuations. In *Between Bands and States*, edited by S. A. Gregg, pp. 135–158. Occasional Paper No. 9. Center for Archaeological Investigations, Southern Illinois University, Carbondale.

Marcus, J., and Flannery, K. V., 1996, *Zapotec Civilization: How Urban Society Evolved in Mexico's Oaxaca Valley*. Thames and Hudson, London.

Meggers, B. J., 1954, Environmental Limitations on the Development of Culture. *American Anthropologist* 56:801–824.

Peebles, C. S., and Kus, S. M., 1977, Some Archaeological Correlates of Ranked Societies. *American Antiquity* 42:421–448.

Price, T. D., and Brown, J. A. (editors), 1985, *Prehistoric Hunter-Gatherers: The Emergence of Cultural Complexity*. Academic Press, San Diego.

Sahlins, M. D., 1963, Poor Man, Rich Man, Big-Man, Chief: Political Types in Melanesia and Polynesia. *Comparative Studies in Sociology and History* 5:285–303.

Sanders, W. T. and Webster, D., 1978, Unilinealism, Multilinealism, and the Evolution of Complex Societies. In *Social Archeology: Beyond Subsistence and Dating*, edited by C. L. Redman, M. J. Berman, E. V. Curtin, W. T. Langhorne, Jr., N. M. Versaggi, and J. C. Wanser, pp. 249–302. Academic Press, New York.

Sebastian, L., 1991, Sociopolitical Complexity and the Chaco System. In *Chaco and Hohokam: Prehistoric Regional Systems in the American Southwest*, edited by P. L. Crown and W. J. Judge, pp. 109-134. School of American Research Press, Santa Fe.

Spencer, C. S., 1993, Human Agency, Biased Transmission, and the Cultural Evolution of Chiefly Authority. *Journal of Anthropological Archaeology* 12:41–74.

Spencer, C. S., 1994, Factional Ascendance, Dimensions of Leadership, and the Development of Centralized Authority. In *Factional Competition and Political Development in the New World*, edited by E. M. Brumfiel and J. W. Fox, pp. 31–43. Cambridge University Press, Cambridge.

Steponaitis, V. P., 1983, *Ceramics, Chronology, and Community Patterns: An Archaeological Study at Moundville*. Academic Press, New York.

Steponaitis, V. P., 1991, Contrasting Patterns of Mississippian Development. In *Chiefdoms: Power, Economy, and Ideology*, edited by T. Earle, pp. 193–228. Cambridge University Press, Cambridge.

Chapter 8

Modeling Tributary Economies and Hierarchical Polities

A Prologue

HENRY T. WRIGHT

INTRODUCTION

Archaeological anthropologists increasingly face, in all our endeavors, the problem of expressing and evaluating our ideas about cultural phenomena, in which participants deal with many impacts from parallel and higher-order entities through the recursive—that is, self-redefining—mechanisms that culture allows. New formal models involving "cultural algorithms" (Reynolds, 1994) provide promising approaches to such problems. Kent Flannery's magisterial study of early foragers at the cave of Guilá Naquitz in the Valley of Oaxaca in Mexico (Flannery, 1986) serves as an inspiration to those seeking to advance theoretical formulations about complex productive and social arrangements.

Anthropologists long ago recognized a class of economic systems, typical of prestate and early state polities, in which goods were not moved from producers to consumers by a direct exchange process. Instead, goods were extracted from producing households by periodic tribute demands, stored, and then redistributed to consumers at the discretion of political leaders. In the 1950s and 1960s, this system was thought to be an equitable distribution process under the charge of benign managerial leaders, and this concept was emphasized in the term *redistributional economy* (Service, 1962:144).

HENRY T. WRIGHT • Museum of Anthropology, University of Michigan, Ann Arbor, Michigan 48109.

Cultural Evolution: Contemporary Viewpoints, edited by Gary M. Feinman and Linda Manzanilla. Kluwer Academic/Plenum Publishers, 2000.

However, in the early 1970s, archaeologists argued that the historically documented examples of such economies were operated by leaders primarily for the benefit of their own apparatus (Earle, 1977; Peebles and Kus, 1977), that is, for the support of spouses, offspring, other relatives, bodyguards, priests, or craft specialists with food and consumer goods and for the suborning of potential competitors with gifts. This realization has been emphasized in the term *tributary economy*. These economies seemed to have had a logic different from that of classic economies. However, none of the so-called economic substantivists in anthropology tried to formalize this logic in models elucidating their structure or operation.

MODELING TRIBUTARY ECONOMIES

The first breakthrough toward the formal understanding of tributary economies was made by someone interested in a completely different problem. In the 1970s, there was much discussion of the role of "population pressure," the approach of human populations to the limits of available resources. Scholars proposed that this pressure motivated innovations in production, such as the beginning of agriculture (Binford, 1968) and the rise of hierarchical polities (Carniero, 1970). Elizabeth Brumfiel (1972) reasoned that if there was increasing pressure on resources before such transformations, there would be an increasingly strong linear correlation between relative estimates of population, as indicated by settlement size, and relative proxies for potential agricultural production, as indicated by the potentially cultivable area around each site. She used settlement-pattern survey information for sites of the prestate Formative periods (700 B.C.–A.D. 100) in the Basin of Mexico, the best such data then available for a developing complex society (Parsons, 1974). She estimated population based on the area of each site in each period and proposed relative potential production from the areas of each soil type in a fixed radius around each site and the known productivity of traditional maize in these soils just after the Mexican Revolution. Surprisingly, Brumfiel found not one linear relation between maize and population, but two (Figure 1). She suggested that this finding indicated that a small set of sites with small territories relative to their populations received maize in tribute from the larger group of sites with larger territories relative to their populations. Brumfiel published these results and went on to make continuing contributions to the study of craft production, organization of urban centers, gender relations, and many other aspects of pre-Columbian central Mexico. For some years, various researchers puzzled over those two linear relationships, trying to convert Brumfiel's relative measures into absolute measures of population and maize production, without useful result.

In the late 1970s, Vincas Steponaitas realized that the structural relation between the production capacities and the populations of tribute-producing villages

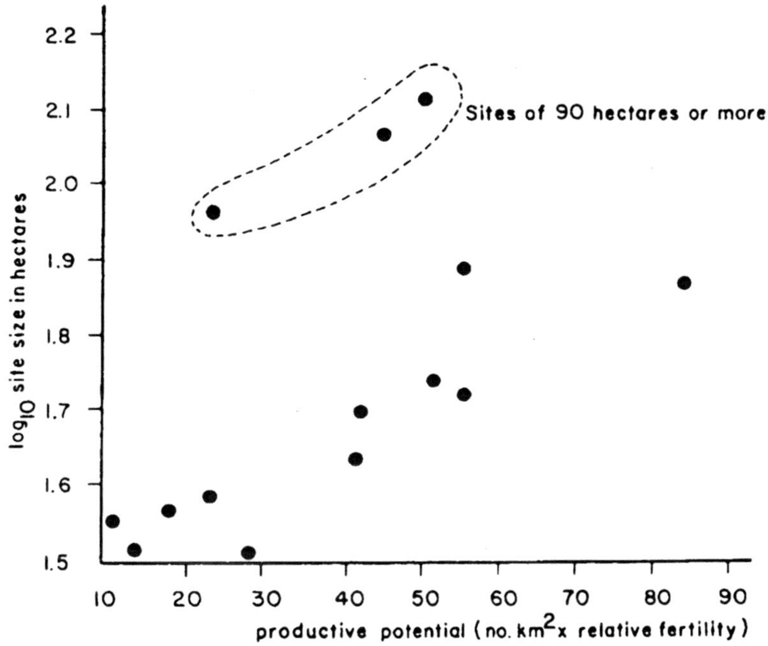

Figure 1. Relative population and productivity of Late Formative centers in the Valley of Mexico (Brumfiel, 1972).

versus tribute-consuming centers could be represented with simple algebraic relations and that these relations produced graphs similar to those Brumfiel had found empirically. Steponaitas's baseline model (Figure 2) is that of villages whose population has grown to the productive limits of their territories or catchments, each of which uses all that it produces to support its own population. The population of each village, V_j, is a function of its annual production, P_j, and a consumption rate per person per year, k.

In a "two-tier" situation (Figure 3) in which each village gives a local center tribute in food, the population of each village, V_{ij}, is a function of its annual production, P_{ij}, and a consumption rate per person per year, k, diminished by the tribute rate demanded of the villages, t_1. In contrast, the population of each local center, L_i, is a function of the sum of (1) the annual production of its own population from its own territory, P_i, and (2) the agricultural tribute it receives from all its subordinate villages, which are a function of the consumption rate per person per year, k, and the portion of each village's product, P_{ij}, taken at the tribute rate demanded of the villages, t_1. Similar algebraic expressions can be devised for a "three-tier" situation in which the local centers are subordinate to a regional center (Figure 4), and beyond.

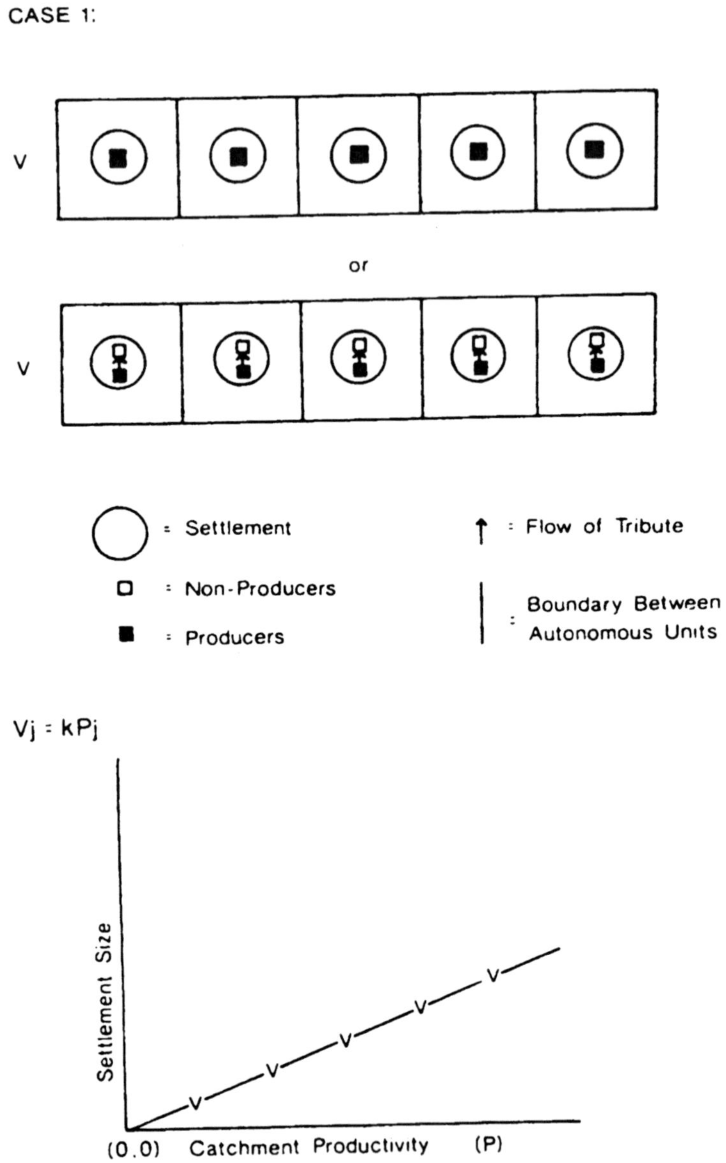

Figure 2. Relative population and productivity with no tribute (Steponaitas, 1981).

Case 2:

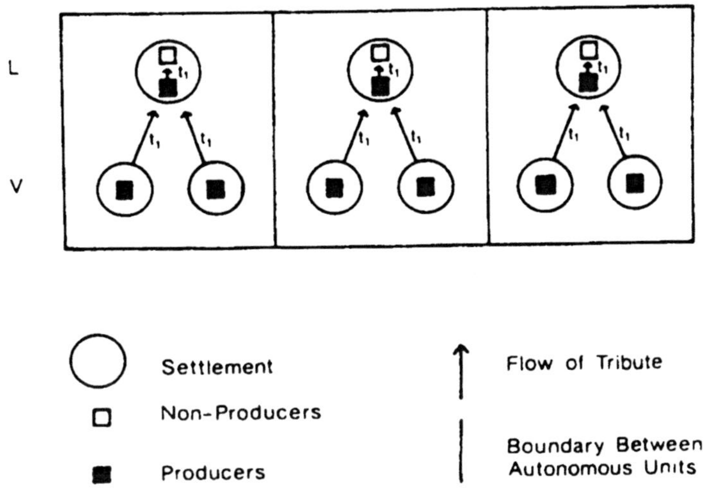

$$L_i = kP_i + kt_1\Sigma P_{ij}$$
$$V_{ij} = k\,(1 - t_1)\,P_{ij}$$

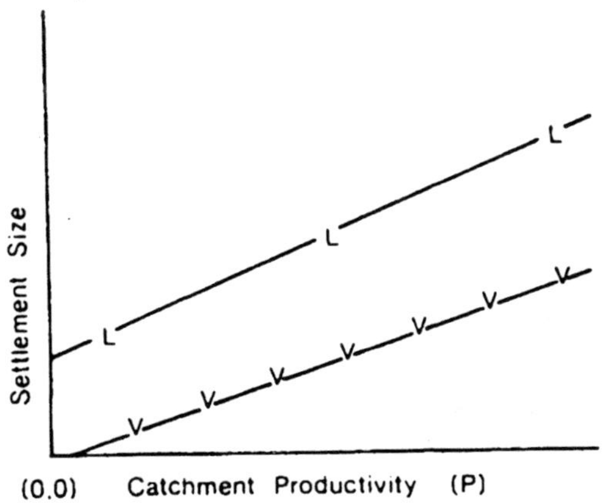

Figure 3. Relative population and productivity with one level of tribute exaction (Steponaitas, 1981).

Case 3:

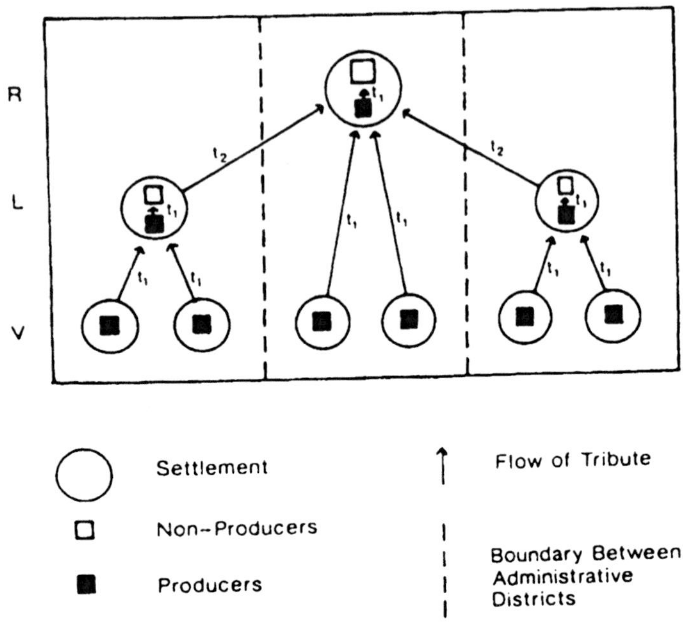

$$R = kP_r + kt_1\Sigma P_{rj} + kt_1t_2\sum^{I}(P_i + \sum^{J} P_{ij})$$

$$L_i = k(1 - t_1t_2)P_i + k(t_1 - t_1t_2)\sum^{J} P_{ij}$$

$$V_{ij} = k(1 - t_1)P_{ij}$$

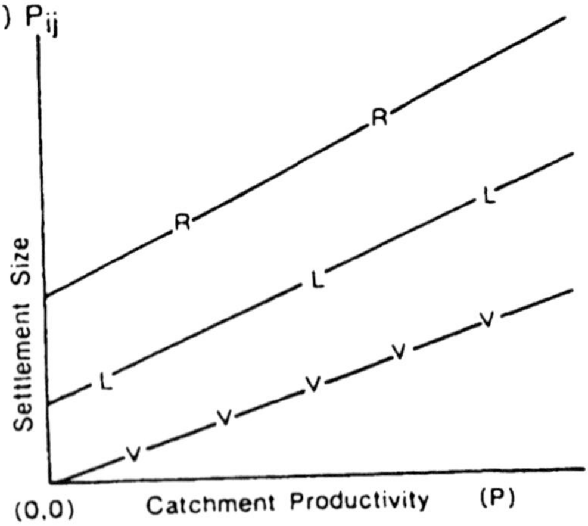

Figure 4. Relative population and productivity with two levels of tribute exaction (Steponaitas, 1981).

EVALUATION OF THE MODELS

Steponaitas evaluated his model with a reanalysis of archaeological survey data from the Basin of Mexico (Steponaitas, 1981). The results conform strikingly to the predictions of the model (Figure 5 is one example) and to the estimated tribute rates reported ethnohistorically for similar political and economic systems. Steponaitas's article drew diverse comment, mostly directed at the specifics of his equations and at empirical measurement problems. Steponaitas turned to major studies of Mississippian chiefdoms (1998). A few researchers tried to apply his model to other cases in other parts of the world. Where adequate survey data and productivity information were available, the results were sometimes promising.

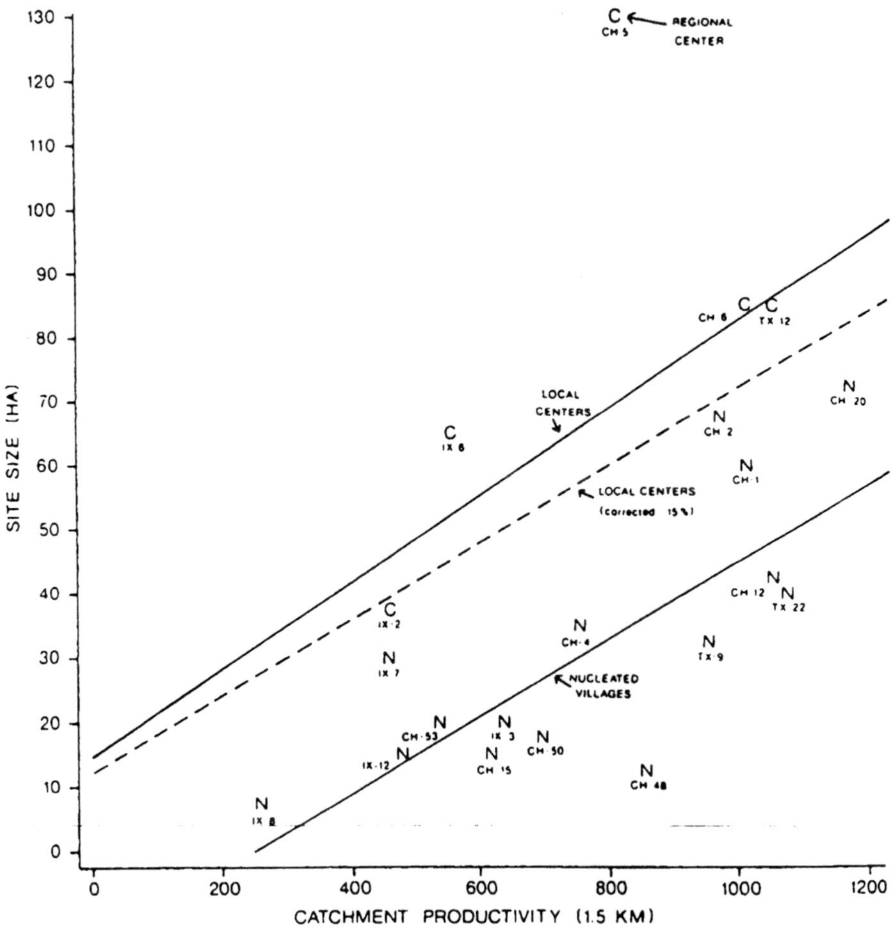

Figure 5. Relative population and productivity of Late Formative settlements in the Valley of Mexico (Steponaitas, 1981).

For example, in the irrigated rice-growing regions of central Madagascar, in the Manandona Valley, an intensive regional archaeological survey was completed by Victor Raharijaona (1986). In the 18th century, sites were surrounded by ditches, and site size and population can be estimated with a precision rare in archaeology (Schwartz, 1992). The paddy area available to each village can be estimated by dividing the nearby irrigable river bottom equally between adjacent sites. Because still-existing rice varieties were grown with methods similar to those used today, Schwartz was able to use modern rice productivities to estimate the maximum rice yield usual for each village. For a restricted area of valley near Sahanivotry, the villages were small and evenly spaced (Figure 6). The rice needed

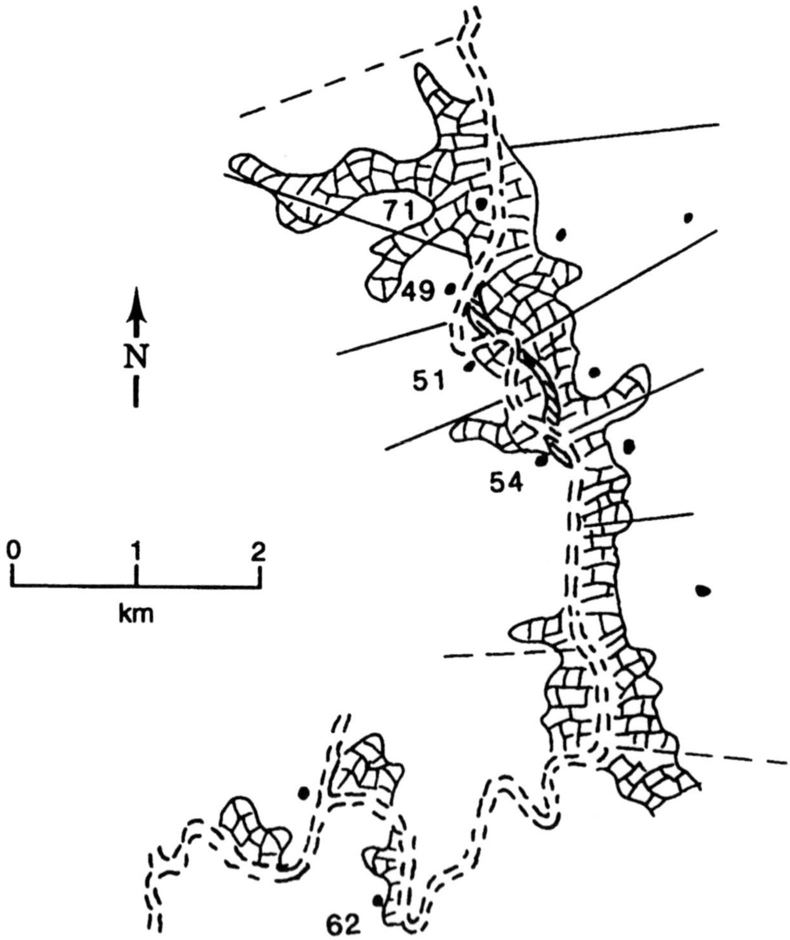

Figure 6. Eighteenth-century settlement and rice paddies near Sahanivotry (Schwartz, 1992).

to support each village approaches the limits of the productivity of available fields, but no village had to obtain food from elsewhere (Figure 7). For a more expansive area of the valley near Manandona, there were more villages and a few larger centers (Figure 8). Some larger centers seemed not have had land sufficient to produce the rice needed (Figure 9). When Raharijaona (personal communication, 1986) asked a noble elder of a still-inhabited center (Site 19 on Figure 8) how they subsisted, the elder explained that his family had a "special relationship" with the adjacent smaller villages, which gave them rice as rent or tribute.

In other parts of the world, although there was other evidence of tributary exaction, the predicted relationship was not found. For example, on the Deh Luran Plain of lowland southwestern Iran, stratified prestate societies with centers and villages predominated during the Farukh phases of the mid-fifth millennium B.C. (Figure 10) (Neely and Wright, 1994). Settlement disposition, large storehouses in centers, and other evidence indicate that elite families exacted tribute (Wright, 1994). However, most sites, and especially the large regional center at Musiyan (Site 20 on Figure 10), had more than enough land to produce the grains such as wheat and barley that the local inhabitants would have consumed (Figure 11). I thought of several sources of error, such as misapprehension of the amounts of land actually cultivable in the past or the importance of subsistence from sheep and goat herds, whose size would depend on the availability of summer pasture in

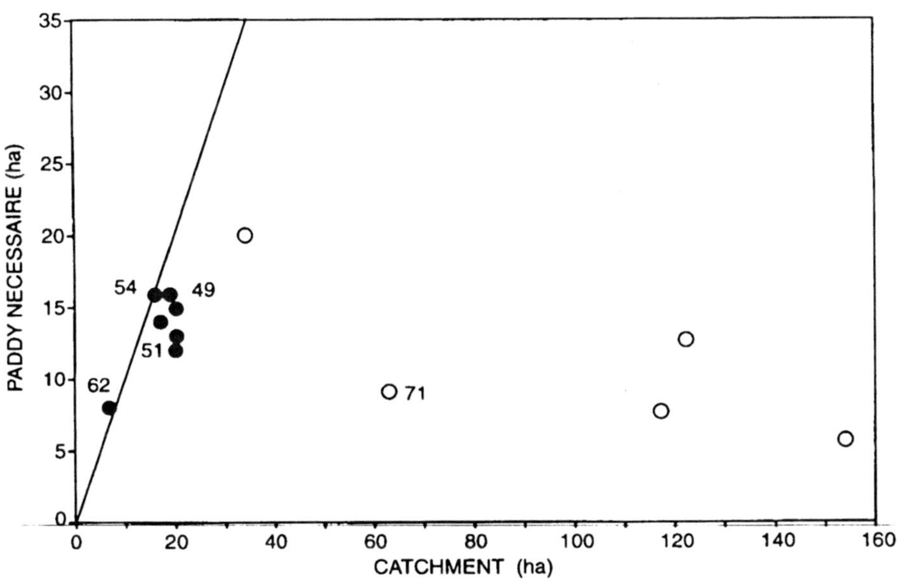

Figure 7. Estimated populations and productivities of villages near Sahanivotry (Schwartz, 1992) (Numbers refer to sites shown in Figure 6).

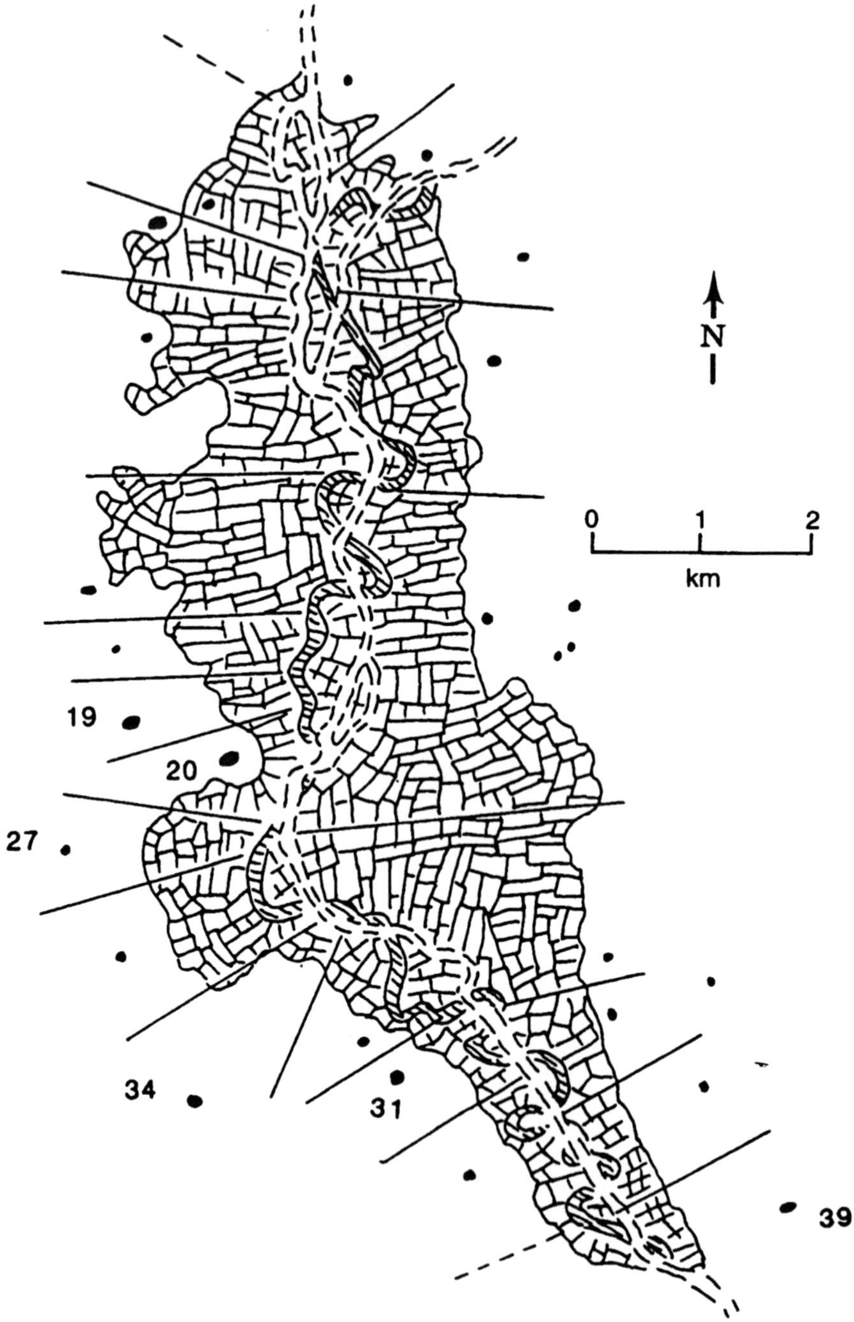

Figure 8. Eighteenth-century settlement and rice paddies near Manandona (Schwartz, 1992).

the mountains. However, Steponaitas suggested a much more interesting possibility. Perhaps, rather than demanding part of each village's harvest, the lords of Musiyan demanded a fraction of each village's labor, and these workers went to the centers and grew crops in land around each center. Steponaitas suggested that such a labor-tribute system would have a graph very different from a goods-tribute system.

GOODS-TRIBUTE VERSUS LABOR-TRIBUTE MODELS

Let us contrast goods-tribute and labor-tribute models of comparable production intensity and hierarchical scale. For consumption and production, let us assume systems where half a hectare of land supports an adult and one adult can cultivate 2 ha of land.

For a situation with tribute in goods (Figure 12), let us assume a three-tier system with (1) a regional center with 5,000 inhabitants, of whom 2,000 are not supportable by the resources of the local territory; (2) two local centers with populations of 3,800 and 3,200 people, of whom respectively 1,000 and 900 are not supportable by the resources of the local territory; and (3) a number of smaller villages, that must provide the goods required to support the additional populations in the centers. The populations of villages are diminished by the 3,900 people in the centers for which their lands must provide food in the form of tribute in

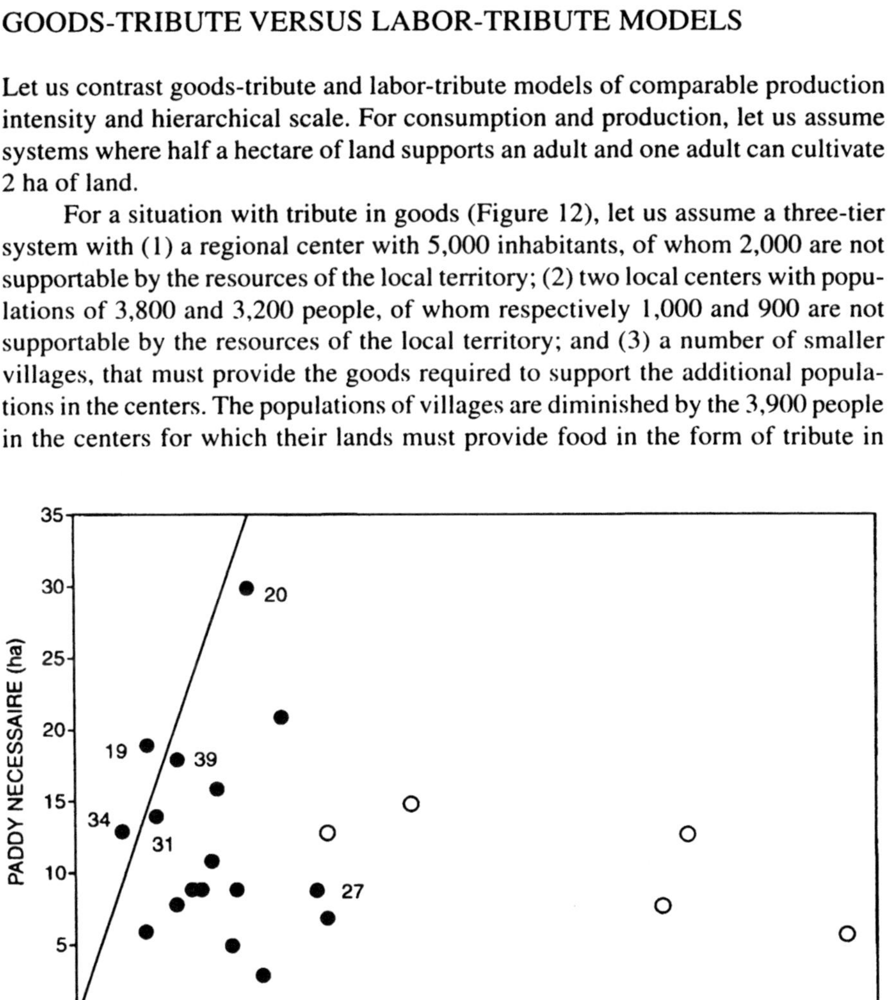

Figure 9. Populations and productivities of villages near Manandona (Schwartz, 1992). (Numbers refer to sites shown in Figure 8).

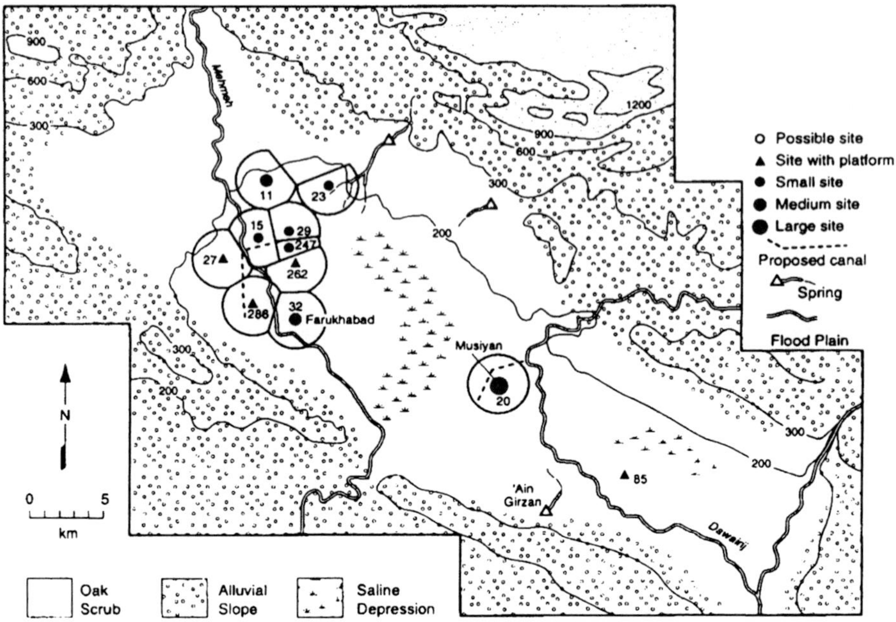

Figure 10. Settlement during the Farukh phase of the mid-fifth millennium B.C. on the Deh Luran Plain (adapted from Neely and Wright, 1994:Fig. V.6).

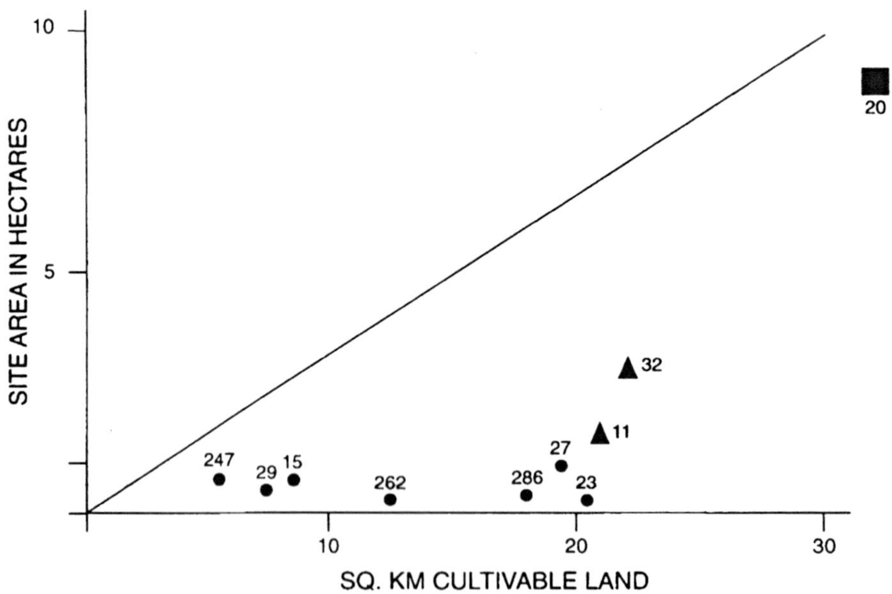

Figure 11. Relative population and productivity on the Deh Luran Plain during the Farukh phase.

goods. For this model, it is assumed that the costs of transport of food from the villages to the centers are negligible. This goods-tribute model produces a graph similar to that found by Brumfiel (Figure 1) and treated analytically by Steponaitas (Figures 2–5).

For a situation with the tribute in labor (Figure 13), the size of villages is unchanged because the inhabitants, who must go periodically to the centers and labor to grow food to support people there who are not supported by their own local laborers, must maintain village residences. It is assumed that this is also a three-tier system with (1) a regional center with 5,000 inhabitants, of whom 2,000 are not supported by local laborers; (2) two local centers with populations of 3,800 and 3,200 people, of whom respectively 1,000 and 900 are not supported by local laborers; and (3) a number of smaller villages that must provide the labor required to support the nonworking elite. It is assumed that the costs of transport of laborers from the villages to the centers are negligible. However, while these laborers are at the centers, which is assumed to be 20% of their working year, they also must produce additional food to feed themselves.

In the exemplary model, the land cultivated around the larger center must make up the 1,500 ha cultivated by local people and the 1,000 ha cultivated by

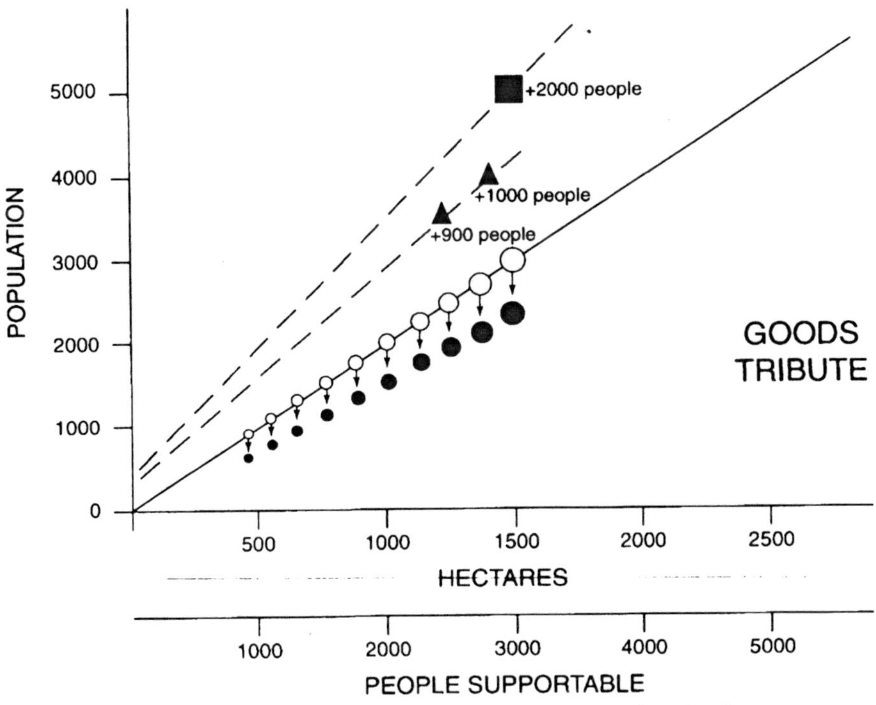

Figure 12. Population and productivity with two levels of goods tribute.

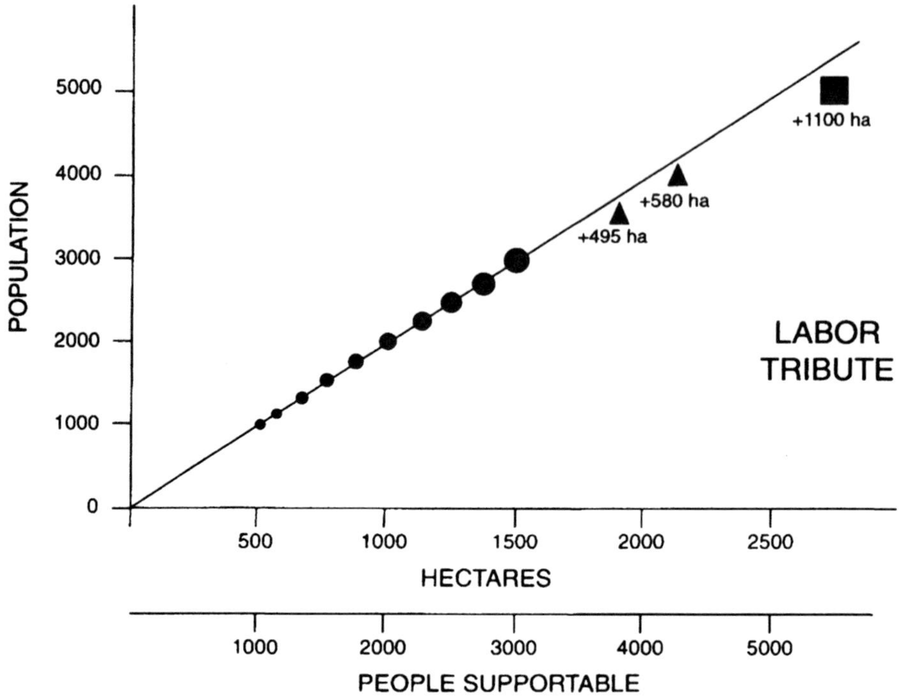

Figure 13. Population and productivity with two levels of labor tribute.

tribute laborers to feed the resident population, plus another 100 ha to feed the visiting laborers.

Similarly, the land cultivated around the larger local center must be 1,400 ha cultivated by local laborers and 500 ha cultivated by tribute laborers to feed the resident population of 3,800, plus another 50 ha to feed visiting laborers.

Finally, the land cultivated around the smaller local center must be 1,600 ha cultivated by local laborers and 450 ha cultivated by tribute laborers to feed the resident population of 3,200, plus another 45 ha to feed visiting laborers.

Obviously, different assumptions about the proportions of nonworking elite members and different allowances for support of tributary laborers generate different graphs. The figures assumed here facilitate comparison between the consequences for settlement hierarchy between the goods-tribute and labor-tribute situations. This labor-tribute model generates a settlement size distribution similar to that noted for the Deh Luran Plain during the Farukh phase in that centers are smaller than their catchments could support (Figure 11). Villages do not approach what their catchments could support, but this is not a necessary feature of the model.

DISCUSSION AND CONCLUSION

The existence of these two types of tributary economic strategies—goods exaction and labor exaction—raises a complicated problem: Under what conditions might a goods-tribute policy or a labor-tribute policy, or even some combination of the two, be used? Criteria for choosing one or the other strategy can be proposed. For example, where land is available and transport difficult, labor tribute might be more effective, but where land is scarce and transport easy, goods tribute might be more effective. However, many, if not most, polities with tributary economies likely used a mix of the two strategies. One compelling reason that this would be expected is that some activities, such as public construction, are probably more efficiently undertaken with labor tribute, whereas others, such as the acquisition of a storable, high-quality foodstuff, are usually more efficiently undertaken with goods tribute.

Another problem not resolved by these rather static structural models of the relation among production, population, elite structure, and tribute is that productivity from year to year is variable and difficult to predict. Political leaders may require and expect a certain amount of tribute, but if production fails, the projected tribute does not materialize. Prudent leaders keep resources in long-term storage to use or redistribute if such shortfalls occur.

These latter problems may best be resolved with an "agent-based modeling" or "artificial society" approach (Epstein and Axtell, 1996; Gilbert and Conte, 1995). In anthropology, such approaches have hitherto been used to understand forager societies and agricultural developments documented by the Oaxaca project of Kent Flannery (Reynolds, 1986), the dynamics of autonomous village societies (Dean et al., 2000; Kohler and Carr, 1997; Kohler et al., 2000) and the decisions of farmers in stratified societies (Lansing, 1991). Such an approach to the operation of hierarchical political economic systems has various advantages. The agents would be political leaders—chiefs and subchiefs—who, given the extant populations and available resources, would decide on their own mixes of strategies to undertake a variety of activities. Political leaders whose strategies were more successful would prosper, and their strategies might be emulated by prudent adversaries. The simulation could be run by using various decision-making rules and rules for the promotion of leaders; predicted settlement patterns, elite residential areas, storage features, and other archaeological evidence could be compared with actual settlement patterns, elite residential areas, storage features, and so on. Also, an agent-based approach to this small and rather structured problem could be a first step toward more global simulations of prestate cultural systems. The activities of agents could be expanded to include additional political negotiations, decisions, and actions. Eventually, networks of competing polities, each with its own internally competing elite, could be generated.

In sum, this chapter has reviewed the development of formal models of tributary economies. It has documented progress in developing testable formal models for cases in which political leaders exact goods from producers, and it has proposed a parallel understanding for cases in which political leaders exact labor from producers. Such relatively simple static models cannot represent more dynamic situations in which a mix of strategies is used in changing environments. Agent-based modeling is proposed as a promising means to represent and address these more complex situations.

ACKNOWLEDGMENTS

The work of Elizabeth Brumfiel and Vincas Steponaitas has been an inspiration over many years. Victor Raharijaona generously introduced me to the Manandona Valley and made some of his data available to me and George Schwartz, who made the population and field area estimates in an unpublished paper. James Neely gave me access to the Deh Luran archaeological survey data. An earlier version of this chapter was given at the Santa Fe Institute in June 1996 at the invitation of George Gumerman and Tim Kohler. Their comments and those of Marcus Feldman were enlightening. Gary Feinman and Eric Rupley pointed our some important weaknesses in the original formulation. I am indebted to all these and many more, but take full responsibility for my uses of their concepts and evidence.

REFERENCES

Binford, L. R., 1968, Post-Pleistocene Adaptations, In *New Perspectives in Archeology*, edited by S. R. Binford and L. R. Binford, pp 313–341. Aldine, Chicago.

Brumfiel, E., 1972, Regional Growth in the Eastern Valley of Mexico: A Test of the "Population Pressure" Hypothesis. In *The Early Mesoamerican Village*, edited by K. V. Flannery, pp. 234–249. Academic Press, New York.

Carniero, R., 1970, The Origin of the State. *Science* 169:733–738.

Dean, J. S., Gumerman, G. J., Epstein, J. M., Axtell, R., Swedlund, A. C., Parker, M. T., and McCarroll, S., 2000, Understanding Anasazi Culture Change through Agent-based Modeling. In *Dynamics in Human and Primate Societies: Agent-Based Modeling of Social and Spatial Processes*, edited by T. Kohler and G. J. Gumerman. Oxford University Press, London.

Earle, T. K., 1977, *Economic and Social Organization of a Complex Chiefdom*. Anthropological Papers No. 63. Museum of Anthropology, University of Michigan, Ann Arbor.

Epstein, J. M., and Axtell, R. L., 1996, *Growing Artificial Societies: Social Science from the Bottom Up*. MIT Press, Cambridge, MA, and Brookings Institution, Washington, D.C.

Flannery, K. V. (editor), 1986, *Guilá Naquitz: Archaic Foraging and Early Agriculture in Oaxaca, Mexico*. Academic Press, Orlando.

Gilbert, N., and Conte, R., 1995, *Artificial Societies: The Computer Simulation of Social Phenomena*. UCL Press, London.

Kohler, T. A., and Carr, E., 1997, Swarm-Based Modeling of Prehistoric Settlement Systems in Southwestern North America. In *Proceedings of Colloquium II, UISPP 13th Congress, Forli, Italy, September 1996*, edited by I. Johnson and M. North. Sydney University Archaeological

Methods Series No. 5. Department of Archaeology, University of Sydney, Sydney, Australia. [http://www.archaeology.usyd.edu.au/resources/documents/ kohler]

Kohler, T. A., Kresl, J., Van West, C., Carr, E., and Wilshusen, R. H., 2000, Be There Then: A Modeling Approach to Settlement Determinants and Spatial Efficiency among Late Ancestral Pueblo Populations of the Mesa Verde Region, U.S. Southwest. In *Dynamics in Human and Primate Societies: Agent-Based Modeling of Social and Spatial Processes*, edited by T. A. Kohler and G. J. Gumerman. Oxford University Press, London.

Lansing, J. S., 1991, *Priests and Programmers: Technologies of Power in the Engineered landscapes of Bali*. Princeton University Press, Princeton.

Neely, J. A., and Wright, H. T., 1994, *Early Settlement and Irrigation on the Deh Luran Plain: Village and Early State Societies in Southwestern Iran*. Technical Report No. 26. Museum of Anthropology, University of Michigan, Ann Arbor. [Summarized at http:// www.umma.lsa.umich.edu/oldworld/ Deh_Luran/Deh_Luran.html]

Parsons, J., 1974, The Development of a Prehistoric Complex Society: A Regional Perspective from the Valley of Mexico. *Journal of Field Archaeology* 1:81–108.

Peebles, C., and Kus, S., 1977, Some Archaeological Correlates of Ranked Societies. *American Antiquity* 42:421–448.

Raharijaona, V., 1986, Reconnaissance archéologique dans le Manandona (Vakinankaratra). *Taloha* 10:73–114.

Reynolds, R. G., 1986, An Adaptive Computer Model for the Evolution of Plant Collecting and Early Agriculture in the Eastern Valley of Oaxaca, Mexico. In *Guilá Naquitz: Archaic Foraging and Early Agriculture in Oaxaca, Mexico*, edited by K. V. Flannery, pp 439–500. Academic Press, New York.

Reynolds, R. G., 1994, An Introduction to Cultural Algorithms. In *Proceedings of the 3rd Annual Conference on Evolutionary Programming*, edited by A. V. Sebald and L. J. Fogel, pp. 131–139. World Scientific Publishing, River Edge, NJ.

Schwartz, G., 1992, Rice and Population in Eighteenth Century Madagascar, Analyzing Archaeological Evidence of the Subsistence Economy on the Central Highlands. Unpublished manuscript, Museum of Anthropology, University of Michigan, Ann Arbor.

Service, E., 1962, *Primitive Social Organization*. Random House, New York.

Steponaitas, V., 1978, Location Theory and Complex Chiefdoms. In *Mississippian Settlement Systems*, edited by B. Smith, pp. 417–453. Academic Press, New York.

Steponaitas, V., 1981, Settlement Hierarchies and Political Complexity. *American Anthropologist* 83:320–363.

Wright, H. T., 1994, Pre-state Political Formations. In *Chiefdoms and Early States in the Near East*, edited by G. Stein and M. Rothman, pp. 67–84. Prehistory Press, Madison. (Reprinted from *The Evolution of Complex Societies: The Harry Hoijer Lectures for 1982*, edited by T. K. Earle. Undena Press, Malibu, 1984.)

Chapter 9

The Development of Administration from Collective to Centralized Economies in the Mesopotamian World

The Transformation of an Institution from System-Serving to Self-Serving

MARCELLA FRANGIPANE

INTRODUCTION

In his article, "The Cultural Evolution of Civilization" (1972), Flannery proposed a timeless model for analyzing the processes of social change, which he claimed was able to provide instruments for a general interpretation of "universal evolutionary mechanisms and processes" leading to the rise of state societies. Flannery's proposal, like most of his scientific contributions, raised fundamental issues encompassing theoretical/conceptual aspects and methodological considerations of importance. One can criticize the excessive confidence he then placed in the capacity of wholly abstract and mechanistic models to offer a universal explanation of historical processes, typical of systems theory that had become fashionable in

MARCELLA FRANGIPANE • Dipartimento di Scienze Storiche Archeologiche e Antropologiche dell' Antichità, Università di Roma, 00185 Rome, Italy.

Cultural Evolution: Contemporary Viewpoints, edited by Gary M. Feinman and Linda Manzanilla. Kluwer Academic/Plenum Publishers, 2000.

those years. Yet some social interaction mechanisms and the underlying function of certain institutional forms that he had identified when building that model have provided useful analytical tools for investigating substantial aspects of the processes leading to the emergence of the state. Furthermore, the historical concreteness always implicit in the way Flannery properly used the comparative approach when constructing even his most abstract generalizations* made his contribution essential reading for more than one generation of scholars, even those belonging to different theoretical schools.

Two of Flannery's concepts that this chapter refers to are those of "system-serving" versus "self-serving" institutions. Flannery introduced these concepts to emphasize the main sociopolitical function of different categories of institutions. He defined system-serving institutions as "special purpose institutions," in the sense that they perform specific tasks with specific purposes, whereas self-serving institutions operate at a higher level, pursuing "general-purpose" ends. Although the term "system-serving" applied to institutions performing special-purpose tasks is unambiguous in the sense that they work in the interest of the community as a whole, the use of the term "self-serving" for institutions with general-purpose functions has not been explicitly clarified and to a certain extent seems to conflict with the model itself. The word "general" refers to the fact that being at a higher level in the control hierarchies, the institution operates on a more general level that controls the other subsystems; if the model is rigidly applied, this task should also be functional in terms of the whole system, thereby contradicting the definition of self-serving. However, in the examples that Flannery used to illustrate what he considered self-serving institutions, particularly those transformed from system-serving to self-serving through a mechanism that he called "promotion," these institutions clearly coincide with power structures, and the processes leading to their emergence are, in reality, based on competition and contrasts between different sectors of society. Take, for example, the change in the role of the military from a special-purpose institution (to organize wars) to the usurpation of the status of kingship (in lieu of proper royal heirs), marking "the transition from the so-called 'theocratic' first-generation civilizations to their more 'militaristic' successors" (Flannery, 1972:413).

Although the systemic model tends to emphasize the general usefulness of every institution to society by reducing the factors of social conflict to mere imbalances and anomalies that are bound to be absorbed, these factors nevertheless implicitly reemerge in Flannery's examples as a locomotive force of social change. The use of the term "self-serving" for the highest-tier institutions, in my opinion, emphasizes the fact that by becoming structures of power they eventually also,

*In my opinion, a sound comparative method must always be based on a thorough examination of the underlying structural aspects of the societies or processes being compared. In other words, for the comparison to be meaningful, it must not be based on superficial formal elements. This requires a historical process-based study of all the individual component parts being analyzed.

and in some cases above all, fulfill their own needs. This point raises again the crucial issue of the bases of power, its social function, and the mechanisms for its reproduction.

In Flannery's clear analysis, the search for institutional processes and mechanisms (using an approach that emphasizes information processing) thereby becomes an instrument that identifies change in the real function of certain institutions in connection with the socioeconomic and political context that, in turn, has produced them. In many cases, unlike what the evolutionary model of system theory suggests, instead of finding a transformation of the same institution's function following mechanisms under which individual subsystems evolve autonomously, one finds completely new institutions generated by radically changed socioeconomic and political structures. These institutions may borrow from the earlier ones certain ways of operating and certain procedures that may be formally similar, but they do not necessarily imply any direct historical derivation from them as evolved forms of the same institutions.

One case that illustrates this situation is the development of the institutions for administrative control over the redistribution of goods in ancient Near Eastern communities. The concepts of system-serving versus self-serving institutions, as we have seen them emerging between the lines of Flannery's article, are powerful analytical tools for highlighting a key change in social function, even though the specific procedures involved were virtually identical. More specifically, I focus on a changeover in administrative institutions from economic environments using collective administration of staple products during the Neolithic to the centralized management of products in fourth-millennium B.C. societies. It is my contention that the two societies in which the administrative institution first originated and then later developed did not represent a single evolutionary path but, on the contrary, two alternative models with wholly different roots, traditions, and developmental trajectories.

REDISTRIBUTION: AN ECONOMIC CATEGORY OR A USELESS FORMAL CONCEPT

In the Near East, the widespread attention given in recent years to administrative material and particularly to clay sealings, because of the new awareness of their interpretative potential,* has shown the significance of these objects as archaeological correlates of redistribution. However, the identification of the use of clay sealings in different socioeconomic and cultural contexts, and for a much longer time frame than was thought possible until only a few years ago, suggests the

*This is to a large extent thanks to the functional study methodologies designed by Enrica Fiandra and her colleagues, which have made it possible to produce an enlightening reconstruction of the many administrative functions of clay sealings.

need to reflect on the usefulness of the concept of redistribution as previously defined. There is a clear need to recognize, define, and, if necessary, distinguish different types of redistribution with profoundly differing social, economic, and political implications.

Polanyi's initial definition of the concept of redistribution did not correlate it with the economic systems of which it was the expression. If anything, Polanyi denied that any such correlation existed, except in the sense that this type of exchange was predominant in certain societies. Polanyi's purpose was merely to recognize ways of organizing the economy, which he called "forms of integration," but which in reality described the ways in which goods circulated in a given social system. In this perspective, redistribution was called simply "appropriational movements toward a center and out of it again" (Polanyi, 1957:250). Although Polanyi declared that redistribution depended "upon the presence of some measure of centricity in the group," he also affirmed that it "occurs on all civilizational levels, from the primitive hunting tribe to the vast storage systems of ancient Egypt, Sumeria, Babylonia, or Peru" (Polanyi, 1957:250–256). Even when describing the causes or the stimuli that were to have led to the appearance of this form of economic integration, Polanyi cited factors that, in reality, occur in widely differing situations. For example, the climatic and environmental conditions in large polities stimulating the redistributional function of central institutions to integrate societal segments that are regionally specialized, or the time lag between harvesting and consumption that on the one hand is an extremely widespread feature (in all agricultural societies) on the other may be considered only a rationale for redistribution in egalitarian societies with the collective administration of goods. Indeed, harvest consumption requires regulation by forms of redistribution only when collective storage is required. This pattern occurs when there are communal modes of production or when the community's economic life is in some manner organized collectively. This kind of integration is present, for example, when activities such as pastoralism or hunting, with the need to be temporarily absent from the farmed lands and the base village, make it necessary to entrust the community, through its representatives, with the custody and management of the harvest, even if agriculture is managed by individual domestic units.

The extremely generic nature of Polanyi's concept of redistribution has led to an indiscriminate use of this term, in which the deliberate vagueness of Polanyi's definition has since been lost, increasingly giving way to specific definitions of redistribution, each associated with widely differing socioeconomic systems. This diversified application of the term, in part because of the confusion to which it has given rise, has revealed the need to identify redistribution in terms of an economic category that is useful for describing, analyzing, and specifying types of societies, with particular reference to archaic societies.

This need has been more completely fulfilled by the theoretical model advanced by Service (1962), whose success was also due to its easy applicability to archaeological contexts. Service characterized redistribution as a particular form of centralized management and coordination for the circulation of goods in economies specific to early hierarchical societies. However, by making redistribution an essential feature of the chiefdom, and of the chiefdom as he defined it, which in reality was merely one of many possible forms of prestate hierarchical societies, Service provided an excessively narrow reading of redistribution. He failed to account not only for the variety of chiefdoms, but also for the many possible forms of redistribution and the diverse types of redistributional societies. Service's basically functionalist theory emphasized the general usefulness of the control exercised over the distribution of goods by a centralized leadership, which solved the organizational and security problems of large communities settled on ecologically differing territories. But although this interpretation applies to some situations, other forms falling within the category of redistribution differ considerably from the model suggested by Service. On the one hand, some types of redistribution lack any real central leadership, and the centralization of resources is communal in character; on the other hand, some redistributional organizations do not really redistribute, but mobilize and centralize resources and labor. The latter return only a fraction of these resources to the people in ways that more closely resemble reinvestment for the benefit of the elites than a reallocation of goods (reinvestment in the labor force by distributing rations and remunerations to employees working on behalf of the central institutions).

Several years ago, Earle advocated the need to draw a distinction between different institutional forms of redistribution, of which he identified four (Earle, 1977:214–216). However, his forms differ so widely in their structural features and the degree to which they represent the economic and social relationships of society as a whole that they fail to offer a definition that clarifies the concept of redistribution so as to improve and specify Polanyi's generic formulation with its low explanatory power. Indeed, I believe that none of the four types of redistribution considered by Earle can satisfactorily represent this centralized form of allocation of goods. The first three are not characterized by central leadership, and the fourth indicates no real reallocation of the centralized goods. Even Earle considered the first type (*leveling mechanisms*) to be different because it did not have an institutionalized structure. The second type (*householding*, i.e., the distribution of goods in a domestic unit) worked at such a narrow level and was at the same time so widespread that it did not characterize any particular form of economic integration. Conversely, Earle's third type of redistribution, *share-out*, namely "the allocation of goods produced by cooperative labor to participants," involved neither organized storage nor institutional forms that controlled distribution operations. Both these features, in my opinion, must be considered intrinsic features of redistribution if this term is taken to mean the centrally administered circulation of goods. *Mobilization*, finally, is the only type suggested by Earle that is correlated with centralized leadership and the only one of his forms that really involves the organized central management of goods and activities. However, although mobilization (as defined by Earle) moves resources from the periphery to the center, namely from the population to the elite or to government institutions, the goods returning from the center toward the common populations are, in reality, only a small part of the centralized goods. Most of these goods are reinvested in activities promoted and controlled by the elites themselves to generate further wealth for their own benefit and to reproduce themselves by strengthening hierarchical relationships and hence enhancing their power. Here again, I propose that the term *redistribution* does not really fit.

I propose that *redistribution* is more suitable for describing situations in which the elite play a real redistributional function, putting back into circulation in various forms (ritualized exchange, reallocation of surpluses on particular occasions, building public works) a fairly substantial part of the collected goods. The concept of mobilization formulated by Earle is useful to characterize stratified centralized societies, as in the case of early state systems. However, I do not believe that it can be used generally as the only possible descriptor for the centralized management of communal goods, also applied to organizations that are specific to ranked hierarchical societies. Mobilization is a form of resource and labor appropriation that marked a qualitative change in socioeconomic relations and the rise of the real economic power of the elites.

Bearing this in mind, and drawing on known documentation of prestate and early state societies in the Near East, I propose that classification of so-called redistributional forms should be revised. I suggest a narrow application of the term *redistribution* as a distinctive feature of economic structure, to be used only for certain types of organizations. I am thinking in particular of those systems in which there is not only a movement of goods from the periphery to the center and vice versa, but also an administratively controlled centralized management of these goods. I therefore propose that the following distinctions and terms be applied.

1. *Share-out:* the distribution of goods produced in common by the community, generally without any gap between production and consumption, characteristic of hunter-gatherer societies.
2. *Egalitarian redistribution:* the reallocation of goods held centrally, which may be produced either collectively or by individual domestic units, but which are, at all events, collectively stored. Central storage is a distinctive element, linked to the time lag between production and consumption, and makes it necessary for the operations to be recorded and controlled, usually by people representing the community. Allocation can take place following different criteria (the produce of each single household can be kept separately or pooled), but the producers and the consumers are the same. This form is specific to egalitarian agricultural societies.

3. *Hierarchical redistribution:* control over the circulation of goods by a centralized leadership that gathers those goods (as offerings, taxes, tribute) and redistributes goods and services. The financing of the leaders is limited mainly to providing for their needs and supporting their public functions (political and religious). The producers and the consumers are not one and the same, but those who pay in receive other goods or services in exchange. This model is specific to ranked societies, or those whose leadership is a socially stable elite.
4. *Centralization:* the form that Earle called mobilization, in which goods and labor are centralized and redistribution is, in reality, reinvested to finance the activities of the elites. There is no longer an identity between producers and consumers or reciprocity in giving and receiving. This form is typical of stratified and early state societies.

THE FUNCTION OF ADMINISTRATION IN REDISTRIBUTIONAL AND CENTRALIZED SOCIETIES IN THE NEAR EAST

The Near East has been considered an ideal environment to recognize and study redistributional systems, because this form of allocation of goods indeed seems to have characterized early centralized societies there, especially in the Mesopotamian world. The early definition by Robert McC. Adams of the nature of social and economic relationships in the early urban societies of southern Mesopotamia is based on the assumption that the origin of their hierarchical relationships lay in the need to make different staple products from various "specialized subsistence zones" circulate within the communities of the alluvium. The main function of the dominant urban institutions (temple and/or palace) must have been, in this perspective, to mediate relationships among specialized groups of producers by collecting and redistributing goods through "a form of ritualized interchange" (Adams, 1966:48–51). This definition, which at that time pointed out one of the main characteristics of Mesopotamian archaic societies, however, stressed the formal functionality of redistribution as a useful system for the entire society and failed to consider the changes in its nature and real function as socioeconomic relations changed through time.

The wealth of archaeological data now at our disposal, and particularly the greater number of administrative materials related to the widespread use of administration practices in Near Eastern contexts, makes it possible to recognize not only changing relationships in the development of centralized societies, but also completely different forms of redistribution in presumably egalitarian Neolithic communities.

The Origin of Administrative Procedures in Contexts of "Egalitarian Redistribution"

In her long and thorough research into the significance and function of tokens in the Near East, Schmandt-Besserat identified the presence of such materials in ancient contexts in the Early Neolithic. Although the emergence of tokens in itself did not imply that they should be considered as "administrative tools," the possibility that ways of controlling the management of goods had developed in very ancient times also was suggested by the recovery of seals at sites belonging to the same chronological horizon and even earlier. These finds occurred at sites throughout the Levant and Anatolia, including Jericho, Abu Hureyra, Ras Shamra VC, Hagoshrim, Tell Ramad, Tell Judaidah ('Amuq Phase A), the bottom levels of Mersin, and Çatal Hüyük. In recent years, other contexts in the Levant and northern Syria have yielded impressions of these seals; the earliest (in gypsum) have been found at sites dating back to the beginning of the seventh millennium B.C.* These sites, which include Bouqras on the Middle Euphrates and El Kowm (Akkermans et al., 1983; von Wickede, 1990:42–49, Plates 6–13), illustrate that these early stamps, most of which bore geometric motifs, were used mainly to seal containers, even though some of them may also have been used for decorative purposes. At a number of Neolithic sites, furthermore, the emergence of seals coincided with that of oval clay objects known as "sling missiles," which were found from that time onward, usually in large accumulations, at various sites in the Near East. The frequent association of these so-called sling missiles with seals, as well as the fragile nature of their composition, which made them rather unsuitable for use as missiles, suggests that they may have had an administrative function. They may have been related, for example, to such accounting activities as reckoning or weighing.

The diffusion of seals and other possible instruments for recording and control in a number of Neolithic communities from the seventh millennium B.C. and after indicates the rise of new societal needs and duties together with changes taking place in the organization of the economy. These shifts along with changes in social and political life during that period might be defined as the second phase of the Neolithic revolution.

The first fully sedentary societies in the Levant and eastern Taurus had reached, in the course of the eighth millennium, an extraordinary development in their capacity to bring together people into large, permanently occupied villages, consequently showing increasing complexity in political and social organization. These new organizational forms would certainly have been based on a progressive expansion of the economic role of agriculture as an important source of food, but at the same time they probably still reflected the traditional links of cooperation that had characterized the societies of the end of the Paleolithic era.

*We use here only calibrated dates.

This highly sedentary model (or organization) that had great social cohesion seems to have broken up at the beginning of the seventh millennium in virtually every area, albeit in different ways. This led to a "crisis" that might perhaps be partly interpreted as a growth crisis. We can in fact hypothesize, on the one hand, a possible overexploitation of the potential of a technologically primitive agricultural economy and a growth in population that expanded beyond the organizational capacity of egalitarian and politically poorly structured societies.* On the other hand, a change in intracommunity social and economic relationships brought about by certain characteristics of the same agricultural economy occurred. I suggest that when agriculture became a basic subsistence activity, it stimulated a split of communities into several economic units (or families) that had markedly reduced needs for mutual cooperation, thus causing a crisis, and then a change, in the organizational structures of the societies.

The responses to the crisis of the seventh millennium varied according to region, and new economic and organizational models emerged as a result of it, all based on integrating the agricultural economy with other resources, mainly livestock herding. Throughout the Near East, these changes gave rise to different historical trajectories, each linked to the specific environmental and socioeconomic characteristics of the various regions that compose this macroregion. Whereas in most regions social groups broke up, sometimes coupled with new forms of mobility and seasonal occupation of settlements, the most successful forms of adaptation seem to have been of two types. The first type is the continuation of the Pre-Pottery Neolithic B (PPNB) tradition in the lower Middle Euphrates valley, paradoxically in regions with a more arid climate. In these areas, such sites as Bouqras still show social stability and a fully sedentary life in which an important role was played by household units, judging from the features of domestic architecture† (Akkermans et al., 1981). This pattern seems to have been based on an expanding agricultural economy, which was probably made feasible by experimental forms of irrigation that may have been practiced by using the water of the Euphrates. The second successful adaptation was the development of a form of territorial organization based on the flexible use of resources and on consequent mobility. This lifeway was evidenced in the late PPNB period in eastern Jezira and was probably firmly established during the Early Neolithic throughout the whole Syro-Iraqi Jezira region (Frangipane, 1996:46–68). Indications of these types of lifestyles and subsistence practices are found in prepottery sites such as Magzalia, whose subsistence patterns were probably derived from the Zagros communities where domestication processes took different routes from those of the western area of the Fertile Crescent.‡

*We cannot exclude, as a third factor, new difficulties arising from a change in climatic conditions.

†The houses were built to a highly standardized module, and were spacious and clearly separated from one another, as one finds in the major sites of PPNB (ninth-eighth millennium B.C.).

‡It seems that there was an earlier and major emphasis on animal domestication in the Zagros, whereas the main focus in the Levant was on agriculture (Hole, 1984).

The most significant and widespread evidence for the administrative use of seals comes specifically from a site in western Jezira, Sabi Abyad in the Balikh Valley (Akkermans, 1996). This settlement, a Neolithic village that dates to the late seventh/early sixth millennium B.C., has yielded architectural remains with in situ materials demonstrating the existence of a community with completely different features from those of the Late PPNB societies, much more consistent with the model assumed here to typify the Jezira groups. The buildings make up three large complexes, in addition to what is perhaps the beginning of a fourth. Each of these complexes includes two architecturally and functionally distinct blocks: a smaller area, with evidence of domestic activities and implements, and a large sector for storage. In a number of spaces, perhaps courtyards, one finds circular structures, sometimes divided by straight small walls inside, which seem to have been the predecessors of the Halaf *tholoi*. These round structures must also have been associated in part with storage activities. The whole area unearthed so far seems to have been used mainly for storing community goods and, to a very large extent, harvests. The size of the storage buildings, particularly in comparison with the areas set aside for domestic use, makes it unlikely that they were used for the needs of individual households. Rather, they were probably used for collective storage relating to large sections of the community.

About 300 clay sealings, which constitute the earliest evidence of clay seal impressions, have been found mostly in two concentrations, one in each of the two main architectural complexes with specialized storage sectors. In each case, the sealings were found in two particular rooms, where they were piled up and associated with tokens, "sling missiles," and numerous other objects made of clay and other materials (Akkermans and Verhoeven, 1995; Duistermaat, 1996:365–370). The largest of these concentrations (228 clay sealings, of which 201 were all in the same room) was directly in the storage area in one of the buildings; as for the other, it is unclear whether it was in the storage sector or in immediately adjacent rooms. Neverthless, in neither instance were the clay sealings associated with containers or stored cereals but were piled up in a room specially set aside for keeping them together. A large number of seals (67) were represented on the sealings, making up some 27 iconographic categories (namely, seals bearing similar motifs); not one single stamp was found (Duistermaat, 1996:353–365). The seals had mostly been impressed several times, and each of them was largely or exclusively used in relation to a single building. All of clay was local.

From these data, it is possible to reconstruct a situation in which extended groups or broad sections of the community collectively stored their harvests, which were then periodically redistributed under administrative control to those entitled to them. This control was probably exercised by community appointees, to manage the goods entrusted to them. The procedures followed may have been similar to those reported for a Libyan community dating to the beginning of this century, which was organized along lines similar to what is hypothesized here (Frangipane,

1996:53–64; Hallaq, 1994). The large number of seals can be explained by the fact that they must have been applied by individuals and/or by individual household units with access to these goods, whereas the location of the removed clay sealings in a specific part of the building indicates that they were stored, at least for a certain period. These stored sealings likely served as a kind of receipt and/or as evidence for the performance of the operation. The local origin of the clay used for the sealings removes any doubt that the sealed containers might have come from outside the immediate area.

We are therefore dealing with operations carried out entirely in the storage area, involving broad sections of society—in other words, a real redistribution of goods that was administratively controlled in an egalitarian context in which the institutions, or individuals carrying out the controls, were required mainly to guarantee compliance with the rules and hence ensure the "fair" circulation of goods according to the needs and value system of the society. This circulation operated in the interests of the whole community. As Akkermans and Duistermaat pointed out, the needs that must have led to the collective management of staple products, which were concentrated and redistributed, might have had something to do with the fact that some of the producers were absent for all or part of the year. Their absence from the place in which their products were stored made it necessary to set up specialized sites or areas serving the whole enlarged community and all its members. This situation existed, for example, in the Libyan communities already mentioned (Hallaq, 1994). A similar type of organization in the Neolithic societies of Jezira was probably related to nascent forms of pastoral nomadism, or to integrated economies in which various closely interrelated production components might have given rise either to forms of economic dimorphism or to an alternation of activities by the same social groups. One interesting case here is the older site at Umm Dabaghiyah, which appears to have been devoted almost entirely to specialized storage space (Kirkbride, 1974), whereas the domestic structures are so few and so marginal that the site may have been occupied only by a handful of families who served as custodians of the stores. Large storage buildings, which were out of all proportion to the individual dwelling units, also seem to have been present in the Hassuna levels at Yarim Tepe I (Merpert and Munchaev, 1993:Fig. 6.3).

Consequently, the emergence and spread of the use of seals is attested earlier in northern areas of the Near East, where glyptics reached a high level of development that is clearly evidenced in the later Halaf period. In contrast, seals were adopted by the societies in Lower Mesopotamia only in a later period. It therefore is proposed that seals met administrative needs that had emerged in what were fundamentally egalitarian communities with a collective management of staple products.

The origin of redistribution and the procedures connected with it in the Near East therefore are not linked to any potentially hierarchical socioeconomic systems, as the southern Mesopotamian communities seem to be. In the latter, a structure based on large households, which probably constituted the main production and consumption units, gave rise to domestic types of food storage of the kind one might find, for example, in the Samarra societies.* Even in the case of large stores, such as that in the Ubaid community at Tell Oueili (Huot, 1994:167–169), no evidence has been found of glyptics and administration. One may infer from this that in the system of large families with a highly sedentary agricultural economy, which was both highly productive and prone to the risk of environmental instability, the basically domestic management of the harvests could also be accompanied by procedures for centrally concentrating surpluses, perhaps managed by nascent elites. But the powers vested in these elites probably had to be based on the rights acquired by their status, and hence on an ideological consensus. We might therefore make the following assumptions to interpret the Oueili "granary": (1) It was not a system for the routine distribution of products, but for granting access at particular times to a common surplus, with the consumption of basic staple products being managed by individual families; (2) control over these surpluses, if based on ideological consent and hence on the recognition of unchallenged rights, did not require full administrative recording in this early stage.

The Change of Administration Function in Stratified Societies: From Unequal Household Control of Resources to "Centralization"

The administrative tool was probably acquired by the Ubaid groups and assimilated into the organizational model of southern origin when this model spread throughout the northern areas and was superimposed on the Halaf socioeconomic structures. This superimposition probably created new relationships in which there seem to have been social and economic inequalities, perhaps as a result of the ethnic/cultural dichotomy underlying the new northern Ubaid society (Frangipane, 1996:126–143).

The location of the administrative material found in these contexts reveals a difference in the behavior and the purposes of the control operations in comparison with those found in the earlier Neolithic cases. Seals and sealings occurred mainly in the dwellings, mostly in the large tripartite buildings, suggesting that control over the distribution and consumption of the goods took place in the context of a household-managed economy. The presence in these buildings of both sealings and seals indicates that the possessors of the seals lived in the places where sealing operations took place, unlike the situation found in centralized contexts, whether egalitarian (Neolithic communities) or hierarchical (early state societies), in which most of the withdrawal operations were performed in centralized storage places by people who did not live there.

*The architecture at Tell es Sawwan consisted of large tripartite houses or houses with a T-shaped floor plan, with clear evidence, at least in the latter case, of domestic storage (Breniquet, 1991).

The different archaeological result of the administrative procedures of Ubaid society—dispersion versus concentration of clay sealings and their association with seals—indicates that this institution now had a new social function and was required to satisfy completely different needs from the ones that had given rise to it in the collective economies of the Neolithic communities of Jezira. It is possible to infer that administrative control over the movement of goods in a family economy reflects nascent forms of dependent labor, in which goods were distributed to people outside the household, probably to people working for the family itself. It is no coincidence that in the northern areas at the end of the Ubaid period one finds the first mass-produced bowls, which were subsequently related to the distribution of meals or food rations.

Administrative control therefore was not exercised in the interests and on behalf of the community, but by sections of society that were acquiring economic privileges. The institution no longer worked to regulate access by all to the goods to which people were entitled, but to maintain and consolidate privileged access by some to certain resources, and in particular to the fruits of other people's work.

In a broad sense, the administrative institution in both types of societies may be considered to be system-serving, insofar as it reproduced existing socioeconomic relations. But whereas the reproduction of an egalitarian-collective system is wholly conservative because it must prevent a shift in the balance of relationships of equality, the reproduction of a hierarchical system is inherently and potentially capable of evolving into more hierarchical or stratified organizational forms because it not only conserves, but tends to heighten the differences between societal segments. This later shift is only possible insofar as the institutions regulating the socioeconomic relationships are no longer system-serving, but have become self-serving, operating in favor of individual privileged members of society, to the detriment of others. In this way, by regulating relationships between households, which were probably one-sided, the administrative activities of the northern Ubaid societies eventually favored the consolidation of vested interests that certain sections of society had by then acquired.

Only this crucial passage is able to explain the far-reaching functional transformation of administrative institutions in the early state centralized systems in the second half of the fourth millennium. In Late Uruk societies, the elites seemed to have acquired a new type of power based on their capacity to centralize resources, above all the labor force, which they reinvested in activities of both public interest and their own. From the Uruk pictographic tablets, we can see that control was exercised over a substantial number of craft activities and livestock rearing, particularly sheep. At the same time, the frequent reference to the distribution of barley, wheat, and beer rations suggests a capacity to concentrate agricultural resources, even though it is impossible to say whether any direct control was exercised over the land or agricultural labor (Green and Nissen, 1987; Nissen, 1986).

Once again, we find highly centralized resources and resource distribution, but the goods that were distributed were, in my opinion, only a small percentage of those collected. The bulk of the concentrated resources were used to finance activities that brought benefits to the elites. The case of the Late Uruk societies best exemplifies the type of system that I suggested calling centralization rather than redistribution in the first part of this chapter.

Administrative activity in these societies was once again linked to centralized storage, and procedures were markedly similar to those found in community contexts (such as at Sabi Abyad): large concentrations of clay sealings in the public areas, often set aside in special places not directly connected with the containers, a large number of people placing the seals, and the total absence of actual seals in the places where administrative transactions were carried out. However, along with the enormous growth in the quantity and frequency of the operations, as well as in the number of people involved, there was also a change in the type of social and economic function played by the administrative institution. First, the people who withdrew the goods were no longer common members of the community, as one might assume them to be in collective economies, but officials appointed by the central authority to perform various tasks, vested with different levels of responsibility. The number of seals represented is an indication of the extension of a new class of bureaucrats delegated by the central authority, who therefore worked on behalf of that authority and in the interests of the elites embodying that authority. The officials probably performed different tasks at various levels of responsibility, thereby creating an expanding system of power delegation (Claessen, 1978).

The administrative instrument and the control and recording procedures used in Neolithic societies were based on a communal and redistributive economy that had perfected a wide range of clay-sealing functions (these served as guarantees against infringements, as well as documents for transactions that were performed, and also bore the "signature" of the individual involved). When these technologies were adopted into the economy of the Ubaid families who were probably linked by unequal socioeconomic relations, their social and economic function changed. Furthermore, these techniques offered the emerging elites an extraordinary tool for resource management.

In the later period of economic and political centralization, it was probably the elites who used the clay-sealing control system in their own family environment who then also employed it centrally. By so doing, they gave this technology an enormous impetus, enabling it to achieve its full potential, with the addition of new instruments (spherical bullae, complex tokens, tablets). Evidence of the relation that existed between the adoption of these procedures in the family environment, their particular development in elite contexts, and their transfer with further enhancement into the public early state sphere comes not only from the large concentrations of clay sealings, often associated with mass-produced bowls, in the public buildings in Late Uruk centers, like Uruk itself, Susa, and Arslantepe, but also from the substantial quantities of these materials found in the residential quarter of elites on the site of Jebel Aruda (Frangipane and Palmieri, 1988–89; van Driel and van Driel Murray, 1983). As at Arslantepe (Ferioli and Fiandra, 1983; Frangipane, 1994; Frangipane and Palmieri, 1983:414–448, 1988–89), the clay sealings at Jebel Aruda were found both in relation to containers in rooms (in which withdrawal operations are assumed to have been performed) and stored away in even greater numbers in small spaces where they had probably been deliberately placed after removal for counting purposes (or discarded following inventories).

This setting aside of the sealings, which has to be associated with the large number of operations they performed and the need to account for them, is found both in the Neolithic redistributive contexts and in the early state centralized ones. In both contexts, the procedures bear similarities. Yet the two institutions are profoundly different in certain other respects and have totally opposite economic functions. Neolithic redistribution involved no interference whatsoever by individuals or institutions either in the production of goods or their circulation, which occurred following rules that pursued the common interests. This kind of redistribution was intended only to sustain the system of equal access to resources, and the persons responsible for administrative control had the sole function of guaranteeing that this access was permitted in compliance with proscribed rules.

In centralized systems, the so-called redistribution is a way of mobilizing resources shifted from one social sector to another, not merely managed in terms of the formal and procedural aspects of their movements. The central institutions in this case not only play an administrative role, but they have a substantial control over the circulation of goods, and partly also over their production. One only has to think of the centralization of such primary activities as livestock rearing and the many craft activities. In this situation, the function of the administrative institution changed radically. A new bureaucracy proliferated in the early state systems, becoming too large to be easily controlled by the central authority. The bureaucracy was then probably linked to the sovereign through ties of material and moral indebtedness, receiving gifts and privileges in exchange for services and becoming more and more hierarchical. The increased administrative needs, both in terms of the number of duties and the extension of the territory to be controlled, must therefore have become an instrument for further social and economic stratification and differentiation. The administrative institution went from being system-serving to become, at least partially, self-serving.

CONCLUDING REMARKS

Public storage and administration are criteria by which archaeologists can identify the centralized concentration of resources and their return to circulation following codified and socially controlled procedures and patterns. But they identify only the formal level of this type of organization and cannot describe the underlying economic and social relationships. These relationships, giving shape to the structure of society, are what characterize the function of the institutions (Godelier, 1977) and, at the same time, determine their potential evolutionary features.

All societies are organized institutionally with the main purpose of reproducing the relationships on which they are based. If they are of an egalitarian type, the mechanisms for their reproduction are generally effective because the emergence of any unbalanced relationship would be in contrast with the same economic structure of the society. Changes are therefore very slow, mostly depending on the emergence of new conditions. But where the relationships are hierarchical or contain elements of inequality, even in embryonic form, the stronger members of society tend to control the reproduction processes to ensure that the one-sided relations are perpetuated, thereby not only preserving but actually increasing the elements of inequality. Hierarchical societies therefore contain the potential for change within them. Through mechanisms of apparent consolidation of existing social relationships, institutions that eventually operate in the interests of certain sectors of society—self-serving institutions—can help to strengthen these sectors by changing the very nature of inequality from being essentially social to acquiring an economic basis.

The administrative institution that emerged in the Neolithic communities of the Near East with the communal management of goods did not, in my opinion, have the intrinsic potential to develop into something structurally and functionally different, because it operated in the interests of the system and of maintaining that system. It changed only when it was adopted by a radically different type of society from the one in which it had originated, where the management of goods was no longer collective and the production and consumption of resources were probably based on unequal relationships. Then the administrative institution acquired new evolutionary potential and changed from a leveling mechanism to an instrument that served to maintain and expand the economic advantage of certain sectors of society, and by passing into the public sphere, to govern the economic and political powers of the central authorities, fostering inter alia the emergence of new social figures, such as bureaucrats.

The transformation of the institution, then, was due not to any evolutionary processes inherent in the way it operated, but was the result of changes in the social and economic relationships that the institution itself was required to regulate. Such an interpretation of the crucial change in the function of redistributive institutions, which are both political and economic regulators, contributes to the old theoretical debate on system functionality versus internal conflicts as factors of development by stressing their changing role in different societies.

Some implications about the history of early state structures in the Near East can also be stressed. The rise of the main instrument for central management of resources and labor (administration), which is the basis of these structures, indeed appears not to be the result of a linear evolutionary process that occurred in the Mesopotamian alluvium, but rather the effect of the combination of different traditions and experiences matured within the wider area of Greater Mesopotamia. The role of Ubaid communities in this process appears crucial, in part because they contained the premises of increasing hierarchy in the common perspective of an evolution from chiefdom to state. Their role is even more important because their northern expansion and interaction with the Halaf groups appears to have created conditions that led to the rise of new elements of social and economic inequality as well as to the formation of effective political and administrative instruments that fostered its reproduction and institutionalization.

REFERENCES

Adams, R. Mc., 1966, *The Evolution of Urban Society*. Aldine Publishing, Chicago.

Akkermans, P. A., Fokkens, H., and H. T. Waterbalk, 1981, Stratigraphy, Architecture and Lay-out of Bouqras. In *Préhistoire du Levant*, pp. 485–501. Editions du CNRS, Paris.

Akkermans, P. A., Boerma, J. A. K., Clason, A. T., Hill, S. G., Lohof, E., Le Miere, M., Molgat, G. M. F., Roodenberg, J. J., Waterbolk-Van Rooyen, W., and Van Zeist, W., 1983, Bouqras Revisited: Preliminary Report on a Project in Eastern Syria. *Proceedings of the Prehistoric Society* 49:335–372.

Akkermans, P. M. M. G., and Verhoeven, M., 1995, An Image of Complexity: The Burnt Village at Late Neolithic Sabi Abyad, Syria. *American Journal of Archaeology* 99(1):5–32.

Akkermans, P. M. M. G. (editor), 1996, *Tell Sabi Abyad: The Late Neolithic Settlement*. NINO, Leiden.

Breniquet, C., 1991, Tell es-Sawwan: Réalités et problèmes. *Iraq* 55:75–90.

Claessen, H. J. M., 1978, The Early State: A Structural Approach. In *The Early State*, edited by H. J. M. Claessen and P. Skalnik, pp. 533–593. Mouton, Paris.

Duistermaat, K., 1996, The Seals and Sealings. In *Tell Sabi Abyad: The Late Neolithic Settlement*, edited by P. M. M. G. Akkermans, pp. 339–401. NINO, Leiden.

Earle, T. K., 1977, A Reappraisal of Redistribution: Complex Hawaiian Chiefdoms. In *Exchange System in Prehistory*, edited by T. K. Earle and J. E. Ericson, pp. 213–229. Academic Press, New York.

Ferioli, P., and Fiandra, E., 1983, Clay-Sealings from Arslantepe VI A: Administration and Bureaucracy. *Origini* 12(2):455–509.

Fiandra, E., 1982, L'archeologia dei sistemi economici. *Le Scienze* (Italian edition of Scientific American) 169:102–112.

Flannery, K. V., 1972, The Cultural Evolution of Civilizations. *Annual Review of Ecology and Systematics* 3:399–426.

Frangipane, M., 1994, The Record Function of Clay Sealings in Early Administrative Systems as Seen from Arslantepe-Malatya. In *Archives before Writing*, edited by P. Ferioli, E. Fiandra, G. G. Fissore, and M. Frangipane, pp. 125–136. Scriptorium, Turin.

Frangipane, M., 1996, *La Nascita dello stato nel vicino oriente*. Laterza, Rome-Bari.

~~Frangipane, M., and Palmieri, A., 1983, A Protourban Center of the Late Uruk Period.~~ *Origini* 12(2):287–454.

Frangipane, M., and Palmieri, A., 1988–89, Aspects of Centralization in the Late Uruk Period in Mesopotamian Periphery. *Origini* 14(2):539–560.

Godelier, M., 1977, *Antropologia e marxismo*. Editori Riuniti, Roma (Italian translation from Horizon, trajets marxistes en anthropologie, Paris)

Green, M. W., and Nissen, H. J., 1987, *Zeichenliste der archaischen Texte aus Uruk*. Archaische Texte aus Uruk volume 2. Freie Universität, Berlin.

Hallaq, D. M., 1994, The Stone Tablet Registers (Khâttâtât). In *Archives before Writing*, edited by P. Ferioli, E. Fiandra, G. G. Fissore, and M. Frangipane, pp. 377–393. Scriptorium, Turin.

Hole, F., 1984, A Reassessment of the Neolithic Revolution. *Paléorient* 10(2):49–60.

Huot, J.-L., 1994, *Les premieres villageois de Mésopotamie. Du village à la ville*. Armand Colin, Paris.

Kirkbride, D., 1974, Umm Dabaghiyah: A Trading Outpost? *Iraq* 36:85–92.

Merpert, N. Y., and Munchaev, R. M., 1993, Yarimtepe I. In *Early Stages in the Evolution of Mesopotamian Civilization*, edited by N. Yoffee and J. J. Clark, pp. 73–114. University of Arizona Press, Tucson.

Nissen, H. J., 1986, The Archaic Texts from Uruk. *World Archaeology* 17:317–333.

Polanyi, K., 1957, The Economy as Instituted Process. In *Trade and Market in the Early Empires*, edited by K. Polanyi, C. M. Arensberg, and H. W. Pearson, pp. 243–270. Free Press, New York.

Service, E. R., 1962, *Primitive Social Organization*. Random House, New York.

van Driel, G., and van Driel Murray, C., 1983, Jebel Aruda, the 1982 Season of Excavations: Interim Report. *Akkadica* 33:1–26.

von Wickede, A., 1990, *Prähistorische Stempelglyptik in Vorderasien*. Profil Verlag, Munich.

Part IV

Conclusions

Chapter **10**

A Concluding Perspective on the Theoretical Contributions of Kent V. Flannery

Tenets for the Next Century of U.S. Archaeology

Gary M. Feinman

INTRODUCTION

"In the spring of 1959, as I sat taking notes in a World Prehistory course at the University of Chicago, I was startled to hear Dr. Robert Braidwood say, 'Would Mr. Flannery please stay after class?' My first reaction was to wonder whether I had flunked the midterm so badly that he was going to advise me to try an alternative career. Instead, luckily for me, I was invited to join the Oriental Institute expedition to Iran" (Flannery, 1982a:163).

For James D. Watson (1993), successful careers in science hinge heavily on several "rules of thumb" regarding the kinds of interpersonal relationships that one has with professors, mentors, colleagues, and even academic adversaries. Such personal contacts are perhaps particularly important in archaeology, where teamwork and collaboration are often highly valued and necessary (Willey, 1988:xi) and the research goals and designs generally require multigenerational programs of concerted study. Certainly such personal ties have been important to Kent Flannery, who throughout his career has acknowledged his intellectual roots to

Gary M. Feinman • Department of Anthropology, The Field Museum, Chicago, Illinois 60605

Cultural Evolution: Contemporary Viewpoints, edited by Gary M. Feinman and Linda Manzanilla. Kluwer Academic/Plenum Publishers, 2000.

scholars like Braidwood, Bernal, and MacNeish (Flannery, 1997; Flannery and Marcus, 1983; Marcus and Flannery, 1996).

As in Flannery's vignette above, my path to archaeology also was set by an undergraduate World Prehistory class. In 1969, I entered the University of Michigan intending to study biology, but the specter of the Vietnam War and the resultant student protests started to tilt my interests from molecules to societies. A rather grueling semester of chemistry and Conrad Kottak's captivating Introduction to Anthropology induced me to check what anthropology courses were available in fall 1970. More through scheduling quirks than intellectual design, I enrolled in Flannery's Anthropology 221. Serendipitously, it turned out to be the only time that he taught the class. Amid my exposure to (and concerted effort to distinguish) such engaging places as Tehuacán, Teotihuacan, Tula, Tenochtitlán, Titicaca, and Tiwanaku, I decided to pursue a major in anthropology with a specialization in archaeology. Before my semesters in Ann Arbor were completed, I also was fortunate to take Flannery's specialized offerings in Mesoamerican and Near Eastern archaeology. In the former, I began to develop my enduring fascination with the Valley of Oaxaca and its history, which in turn led me to attend the City University of New York, where I pursued graduate training under Flannery's former student, Richard Blanton. The rest, as they say, "is history," at least my own personal history.

Flannery's insights and teachings have left their indelible mark on my own research. Yet far more significant is the foundation that his cumulative writings have provided for contemporary archaeology as we enter the next century. Flannery's broad intellectual contributions to the discipline are evident in the collected chapters in this volume as well as in the prestigious awards that he has received, which include the American Anthropological Association Distinguished Lecture Award, the Alfred Vincent Kidder Award for Eminence in the Field of American Archaeology, and election to the National Academy of Sciences.

FOUR KEY TENETS FOR CONTEMPORARY ARCHAEOLOGY

In the remainder of this chapter, I discuss four broad tenets and principles—a comparative yet historical approach, holistic analyses that range from the ecological to the ideological, a multiscalar focus, and research designs that examine cultural as opposed to arbitrary/statistical units—that seem to be core features in Flannery's seminal writings. Collectively, these tenets not only have proved influential for contemporary archaeological research as practiced in the United States and elsewhere, but they provide an important theoretical road map for the discipline in the 21st century. Flannery's expression of these four principles was never promoted as a new theoretical dogma or the next "ism"; rather, each tenet was thoughtfully embodied and illustrated through the presentation of his own empirical work.

A Comparative Historical Framework

Beginning with his writings, Flannery has forged a critical intermediate position between noncomparative, highly inductive positions (strict historicism and archempiricism) and the rigid universalizing approaches that search for covering laws, prime movers, and unilinear sequences of change (Flannery, 1967, 1973a, 1986). In the early 1970s, Flannery (1972, 1973b) illustrated that key human transitions, such as the beginnings of agriculture and the emergence of states, are characterized by both broad cross-cultural parallels and equally significant variation in the sequences of historical change and the relative importance of diverse causal factors and connections. Previous comparative efforts had tended to stress similarities while minimizing key elements of variation. By taking a comparative yet still explicitly empirical perspective, Flannery began to describe and distinguish processual commonalities from specific and historically contingent regional differences. Globally, he noted that the historical pathways to sedentary farming villages (from mobile hunting and gathering) or to the rise of states are never strictly alike, but certain notable processes can be seen across different temporal and spatial cases.

In this volume, the cumulated insights of a strongly comparative (yet richly empirical) approach are evidenced in Adams's observations about the emergence of urban societies and technological change as well as in Drennan's theoretical considerations of the distinct conditions and pathways that may lead to the institutionalization of inequality. Yet lurking behind each of the contributions are the core questions about how and why the major shifts in the nature of inequality (Drennan, Spencer and Redmond), the emergence of agricultural and other technological innovations (Smith, Carneiro, Adams), the development of hierarchical political formations (Wright, Frangipane), and the nature of economic specialization (Feinman and Nicholas) were different or similar in particular study domains or historical circumstances as compared with other world regions.

A Holistic Orientation

A second hallmark of Flannery's research has been its explicit intent to integrate the economic and ecological with the social and cognitive aspects of human behavior (e.g., Flannery and Marcus, 1976a; Marcus and Flannery, 1994). The holistic integration of these different cultural spheres might seem to be no more than standard and sensible practice in anthropological archaeology. Yet Flannery's adherence to this balanced research strategy has been maintained over the last three decades in the face of extreme interpretive pendulum swings: first, in the 1960s and 1970s, toward narrow environmental and demographic determinism and, later, in the 1980s and 1990s, toward postprocessual and postmodern approaches that often eschew a direct concern with subsistence, production, and exchange. In this volume, Spencer and Redmond's chapter most directly illustrates the insights gained from this kind of holistic approach. In their chapter, they outline and examine changing patterns of ritual, public architecture, economic production, exchange, and militarism in the Cuicatlán Cañada (Oaxaca, Mexico) during a period when that region seems to have fallen under the domination of the Monte Albán state. By weaving diverse strands of evidence together, Spencer and Redmond provide a cogent and convincing case for extraregional conquest and local change in Cuicatlán.

A Multiscalar Perspective

Few volumes of the 20th century have had more impact in American archaeology than *The Early Mesoamerican Village* (Flannery, 1976). Time has confirmed Wauchope's (1977:656) apt comments in the first paragraph of a book review in *American Antiquity*. "It [the volume] is at least three treats in one package: (1) a significant contribution to one of the least understood topics of Mesoamerican archaeology—the Formative or Preclassic village; (2) an excellent exposition and application of various archaeological analytical methods; and (3) a whale of a lot of fun to read." Yet not until the final paragraph of his review did Wauchope (1977:659) briefly note what I believe may have been one of the most enduring and important features of the volume, the application of an explicitly multiscalar approach to the Mesoamerican Preclassic period, which integrated observations and analyses at a series of nested scales from intrahousehold activity areas through communities and regions to the Mesoamerican macroregion (see also Flannery and Marcus, 1983).

After the publication of *The Early Mesoamerican Village* in 1976, archaeologists continually debated what is the "proper" unit of investigation or the "appropriate" scale of selection (see Feinman, 1996). Yet, whether one looks at the modern world or the ancient world, it is increasingly evident that you cannot understand human organization and behavior by considering any single scalar vantage in isolation (Drennan, this volume). Rather, a multiscalar perspective is needed, and the sad reality for archaeology is that most global regions still cannot match the empirical foundation, scope of coverage, and conceptual universe that Flannery brought to the examination of Preclassic Oaxaca decades ago. In this volume, a number of the chapters (Spencer and Redmond, Feinman and Nicholas) draw explicitly on multiple scales in their examinations of ancient socioeconomic organization and change. Other chapters (Drennan, Wright, Frangipane) conceptualize key processual episodes from several vantages (e.g., leaders, followers, settlements, regions), thereby providing informed interpretations of these key political transitions.

A Focus on Cultural as Opposed to Arbitrary Units

A final distinguishing tenet of Flannery's investigations is their focus on culturally meaningful units as opposed to arbitrary archaeological ones. This feature of his research is evidenced whether he is discussing ritual space and public buildings (Flannery and Marcus, 1976b) or residential contexts (Flannery, 1976), and it is one of the elements that gives Flannery's writing a vital and humanistic character despite its strong empirical and scientific foundation. In *The Early Mesoamerican Village*, Flannery's (1976) emphasis on the examination of culturally meaningful units and broad horizontal uncoverages served as an important alternative to both the traditional reliance on vertical excavation shafts ("telephone booths") and to the contemporary emphasis on frequently scientistic sampling designs and test-pit strategies. The latter strategies and designs certainly have use, especially at early stages of field research. Yet Flannery's eloquent discussions illustrate that in most cases (and for many research problems) the examination of cultural units provides interpretable and anthropologically meaningful results in much the same way that the excavation of natural stratigraphic layers is generally preferable to excavation in arbitrary units.

Flannery's focus on culturally meaningful units also served to stimulate the now-growing emphasis on "household archaeology" (e.g., Manzanilla, 1986; Wilk and Ashmore, 1988). The concern with houses not only provides a direct means for archaeologists to examine questions on production, consumption, and access, but it draws archaeology away from an overriding concern with temples, tombs, museum pieces, and things broadly elite toward a more rounded picture of past societies and everyday life (Adams, this volume).

CONCLUDING THOUGHTS

Over 25 years ago, Flannery (1973a) somewhat boldly predicted the theoretical road that he believed archaeology would travel over the forthcoming decades. His predicted path deviated from both traditional culture history, as practiced during the years preceding, and the "law-and-order" school, which was then in vogue. The latter advocated the search for universal explanations and the adoption of a natural science model for research designs and the assignment of cause and effect. In general, Flannery's prediction was on the mark. A core of archaeologists (a number represented in this collection) have followed a middle-ground theoretical course, having much in common with the path that Flannery outlined in 1973. At the same time, many fewer archaeologists have adhered to a law-and-order approach, retained the search for covering laws, or maintained a faith in the universal appropriateness of linear causality.

Of course, more than 25 years ago, the polar reactions from the postprocessual and archempiricist (Darwinian) paradigms were less explicitly anticipated (Feinman, 1994). Yet this is not the time for those of us who ascribe broadly to the four tenets outlined above to throw in the towel or adopt a new "ism." Flannery (1982b:277–278; Marcus and Flannery, 1996:27–29) has long recognized and forewarned that the kinds of archaeological knowledge required to discuss holistic long-term histories take many decades of concerted and consistent field study and analysis to procure and produce. Frankly, archaeology is a multigenerational endeavor. By now, we must realize that solid accomplishments are often painfully slow. In many global regions, we are just beginning to realize a rich multiscalar empirical record after many decades of study. In some of these areas, the progress that has been made owes much to the broad guiding perspectives laid out by Kent Flannery, his teachers, collaborators, colleagues, and students in the Valley of Oaxaca. May one of Flannery's enduring legacies to the discipline include the steadfast implementation of these four overarching tenets, so amply illustrated through his own research, to present and future archaeological practice in the 21st century.

REFERENCES

Feinman, G. M., 1994, Toward an Archaeology without Polarization: Comments on Contemporary Theory. In *Caciques and Their People: A Volume in Honor of Ronald Spores*, edited by J. Marcus and J. F. Zeitlin, pp. 13–43. Anthropological Papers No. 89. Museum of Anthropology, University of Michigan, Ann Arbor.

Feinman, G. M., 1996, Conclusions. In *Pre-Columbian World Systems*, edited by P. N. Peregrine and G. M. Feinman, pp. 115–121. Prehistory Press, Madison, WI.

Flannery, K. V., 1967, Culture History versus Cultural Process: A Debate in American Archaeology. *Scientific American* 217(2):119–122.

Flannery, K. V., 1972, The Cultural Evolution of Civilizations. *Annual Review of Ecology and Systematics* 3:399–426.

Flannery, K. V., 1973a, Archeology with a Capital S. In *Research and Theory in Current Archeology*, edited by C. L. Redman, pp. 47–53. John Wiley, New York.

Flannery, K. V., 1973b, The Origins of Agriculture. *Annual Review of Anthropology* 2:271–310.

Flannery, K. V., 1982a, Early Pig Domestication in the Fertile Crescent: A Retrospective Look. In *The Hilly Flanks and Beyond: Essays on the Prehistory of Southwestern Iran, Presented to Robert J. Braidwood*, edited by T. C. Young Jr., P. E. L. Smith, and P. Mortensen, pp. 163–188. Studies in Ancient Oriental Civilization No. 36. Oriental Institute, University of Chicago, Chicago.

Flannery, K. V., 1982b, Distinguished Lecture: The Golden Marshalltown: A Parable for the Archeology of the 1980s. *American Anthropologist* 84:265–278.

Flannery, K. V., 1986, A Visit to the Master. In *Guilá Naquitz: Archaic Foraging and Early Agriculture in Oaxaca, Mexico*, edited by K. V. Flannery, pp. 511–519. Academic Press, Orlando.

Flannery, K. V., 1997, In Defense of the Tehuacán Project. *Current Anthropology* 38:660–662.

Flannery, K. V., and Marcus, J., 1976a, Formative Oaxaca and the Zapotec Cosmos. *American Scientist* 64:374–383.

Flannery, K. V., and Marcus, J., 1976b, Evolution of the Public Building in Formative Oaxaca. In *Cultural Change and Continuity*, edited by C. E. Cleland, pp. 205–221. Academic Press, New York.

Flannery, K. V. (editor), 1976, *The Early Mesoamerican Village*. Academic Press, New York.

Flannery, K. V., and Marcus, J. (editors), 1983, *The Cloud People: Divergent Evolution of the Zapotec and Mixtec Civilizations*. Academic Press, New York.

Manzanilla, L. (editor), 1986, *Unidades habitacionales mesoamericanas y sus áreas de actividad.* Serie Antropológica 76. Instituto de Investigaciones Antropológicas, Universidad Nacional Autónoma de México, Mexico.

Marcus, J., and Flannery, K. V., 1994, Ancient Zapotec Ritual and Religion: An Application of the Direct Historical Approach. In *The Ancient Mind: Elements of Cognitive Archaeology*, edited by C. Renfrew and E. B. W. Zubrow, pp. 55–74. Cambridge University Press, Cambridge.

Marcus, J., and Flannery, K. V., 1996, *Zapotec Civilization: How Urban Society Evolved in Mexico's Oaxaca Valley*. Thames and Hudson, London.

Watson, J. D., 1993, Succeeding in Science: Some Rules of Thumb. *Science* 261:1812–1813.

Wauchope, R., 1977, Review of *The Early Mesoamerican Village* (edited by K. V. Flannery, 1976). *American Antiquity* 42:656–659.

Wilk, R. R., and Ashmore, W. (editors), 1988, *Household and Community in the Mesoamerican Past*. University of New Mexico Press, Albuquerque.

Willey, G. R., 1988, *Portraits in American Archaeology: Remembrances of Some Distinguished Americanists*. University of New Mexico Press, Albuquerque.

Postscript

LINDA MANZANILLA

Not many archaeologists have the opportunity to work in several different parts of the world. Of those few who do cross-cultural research, even fewer are interested in also studying general social processes through a careful examination of archaeological data. An even smaller number excel in the theory and methodology of both regional research and activity area analysis. Kent Flannery is one of this select group of archaeologists.

Throughout the years that I have known Flannery, his influence on the type of archaeology that I am interested in has increased logarithmically. Not only have his teachings influenced my work on the development of interdisciplinary methodologies for activity area and domestic archaeological research (i.e., Manzanilla, 1985a, 1986a, 1987a, 1988–1989, 1990, 1993, 1996a; Manzanilla and Barba, 1990; Manzanilla and McClung de Tapia, 1997), but also my comparative research in Mesoamerica (Manzanilla, 1992b, 1995, 1996b, 1997), Mesopotamia (Manzanilla, 1985c, 1986b, 1987b, 1991), Egypt (Manzanilla, 1982, 1985b, 1986c, 1991), and the Andean region (Manzanilla, 1992a).

In his early research in the Near East, Flannery was creative in assessing how broad-range gathering patterns, with a certain degree of seasonal migration from zone to zone and often in altitudinal gradients, provided the background for the domestication of plants and animals. This process involved the movement of wild grains to niches to which they were not adapted, the removal of certain natural selective pressures that allowed mutants to survive, and the selection of nonbeneficial characteristics (Flannery, 1965:3–5; see also Flannery, 1973).

LINDA MANZANILLA • Instituto de Investigaciones Antropológicas, Universidad Nacional Autónoma de México, Cuidad Universitaria, 04510, Mexico City, Mexico.

Cultural Evolution: Contemporary Viewpoints, edited by Gary M. Feinman and Linda Manzanilla. Kluwer Academic/Plenum Publishers, 2000.

With respect to animals, Flannery studied changes in the proportions of sex and age in animal populations in early Near Eastern sites. He established that a tendency to domestication through captivity is indicated when the percentage of butchered young animals is greater than that of adults and when more male than female animals are butchered (Flannery, 1965:7–8).

Guilá Naquitz, a shallow cave in the Valley of Oaxaca, gave Flannery (1986) insight into archaic territorial behavior and resource exploitation. His research on the cave provided detailed information on the scheduling and seasonal-procurement activities of a small foraging population in Oaxaca. Guilá Naquitz was the location of a family microband camp that was part of a macroband foraging system, which also included camps for maguey roasting, chert quarrying, and hunting (Flannery, 1986:41).

Flannery was particularly interested in village life (1972, 1976a). Influenced by a paper presented by Struever in 1969, the now classic methodological perspective outlined in *The Early Mesoamerican Village* (Flannery, 1976a) made explicit the different levels of spatial analysis in archaeology. In this scenario, the "activity area" is the smallest observable unit (Flannery, 1976b:5–6) where debris and artifacts are articulated. Groups of activity areas may be associated with male and female work areas of the household (Flannery, 1976b:5). These areas may be observed through the extensive excavation of complete houses and external areas and the recording of artifacts and debris in situ. The house may be defined by its floor and the limits of construction, and we may investigate whether particular aspects of the construction itself are related to status, specialization, and/or access to foreign materials.

The next level is the "household unit." This unit is composed of the house and all the surrounding features (Flannery, 1983; see also Flannery and Winter, 1976) that may be related to internal and external household activities. Groups of houses around patios or plazas may represent kin groups that share activities. Their dumps may differ with respect to other similar groups, providing the possibility of distinguishing activities shared by all households (universal activities) from those shared by some households in all sites (possible household specialization), those present only in some settlements (possible regional specialization), and those characteristic of only one settlement (unique specialization) (Flannery and Winter, 1976:36). Given an adequate chronological framework, the degree of specialization at the household and village levels may thus be properly addressed.

These last considerations are of the utmost importance to me, particularly when I see many Ph.D. dissertations that consider only surface data. The emphasis on careful extensive excavation of complete sectors of a settlement, particularly primary contexts, together with a regional framework from which to choose adequate settlement types, allows a precision in functional, chronological, and ecofactual data that cannot be gained in any other manner. I would add that circulation and distribution networks may be defined when production and use/consumption sectors are compared.

Additional scales of spatial analysis include the community (see also Flannery, 1972), the settlement and its catchment area, and the region (Flannery, 1976a), where site spacing, relations between soil types and settlement sizes, exchange networks, the development of hierarchies, religious systems (particularly information flow through rituals), and settlement patterns can be discerned. Flannery's interest in systems theory, and in matter, energy, and information flows in ecosystems, drew him to conclude that the complexity of an ecosystem may be inversely proportional to the political complexity of the social system.

In sum, in *The Early Mesoamerican Village*, Flannery (1976a) gives us an example not only of good archaeology, but of how to deal with anthropological interpretations through archaeological data. Flannery's investigations have studied processes that characterize central areas (with long occupational sequences, such as the Valley of Oaxaca) compared with those in marginal areas (such as the Tehuacán Valley). Particularly interesting is the transformation of dispersed domestic units in the Early and Middle Formative to the patio groups of the Late Formative and the domestic compounds that were built around patios in Monte Albán IIIa.

Flannery's research goals at San José Mogote were to assess the origin of sedentary villages, the appearance of social rank and the centralization of power, as well as the process of state formation. The evolution of public structures on stone platforms may be seen as an indicator of political complexity; the centripetal force of San José Mogote with respect to the rest of the valley may also be related to its role as a distributional center, as well as to an increase in craft production and specialization (Flannery and Marcus, 1983). The emergence of this large regional center may be compared to similar processes at other large centers, particularly Chalcatzingo and San Lorenzo Tenochtitlan during the Early and Middle Formative (Flannery, 1982).

The emergence of Monte Albán as the political capital of a three-branch confederation was a further development of the process (Marcus and Flannery, 1996). Indicators of the appearance of a state society include the detection of the first palaces, the first royal tombs, and standardized two-room temples, as well as a settlement hierarchy of three or more tiers (Marcus and Flannery, 1996).

In the remainder of this postscript, I stress Flannery's importance for Mexican archaeology, as well as the new perspectives for the study of large urban settlements that are possible with his methodology (Flannery, 1976a).

For many decades of this century, Mexican archaeology was trapped around monumental archaeological sites. The development of the tourist industry in the country required extensive projects to open sites for visitors. Few archaeologists devoted their research to domestic structures, caves, garbage dumps, or other low-profile sectors of the Mexican territory. The few archaeologists who did were concentrated in the Department of Prehistory at the Instituto Nacional de Antropología e Historia (INAH).

In 1973 Dr. Guillermo Bonfil, at that time the General Director of INAH, proposed to Angel Palerm and José Luis Lorenzo the organization of the First Workshop of Advanced Training in Archaeology. Lasting for three months (from June 12 to September 3, 1973), the goal of the workshop was to train 10 Mexican archaeologists in the key methodological and theoretical aspects of the discipline. These persons would live, eat, discuss, do survey, and excavate with Pedro Armillas, Kent V. Flannery, and William T. Sanders in the Basin of Mexico, the Valley of Oaxaca, and the Maya area.

The majority of these 10 selected archaeologists would either be candidates to head the regional centers of INAH (Otto Schöndube, Lorenzo Ochoa, Arturo Oliveros, Luis Rodrigo, and Emilio Bejarano) or would remain in the different departments of the institution: Salvage Archaeology (Jordi Gussinyer and Alejandro Martínez), the Department of Prehistory (Linda Manzanilla), and the National School of Anthropology and History (Manuel Gándara). Teresa Rojas came from the Centro de Investigaciones Superiores of INAH, and Marcella Frangipane joined the group because she was collaborating with the Department of Prehistory at that time.

It is now 26 years since we experienced this wonderful opportunity of having first-hand interactions with these great scholars on such topics as archaeological survey, landscape study, demography, spatial analysis, activity area research, intensive agriculture, economic symbiosis, and energy flows. This experience certainly left its mark on this group of Mexican scholars, many of whom are now leading figures in their fields.

Particularly in the theoretical part of the workshop, Flannery conveyed his basic concerns about what archaeology can do for anthropology, how archaeology can become a true science with 80–90% probability rates, and how research can be comparable across various regions. Flannery also stressed that ethnographic analogy is limited in scope, in that the different social organizations of the past clearly outnumber those present today. He was opposed to the common assumption that modern environments are comparable to past conditions (he often stated that the modern Arctic tundra was by no means comparable to Paleolithic conditions).

Undoubtedly, this First Workshop of Advanced Training in Archaeology opened new perspectives for research in Mexico. When I think about it, a series of anecdotes comes to mind. At that time I was very critical of how archaeological data were obtained; I was then a student of José Luis Lorenzo, who was influenced by the British environmental school. I did not agree with the presumed relation between microbands or macrobands and hearth size; I also criticized the definition of early ceramic phases based on only a few colluvial eroded sherds.

One of the memorable moments with Kent Flannery in Oaxaca—leaving aside those where tasting different varieties of mescal was the issue—was a particular discussion on typologies (Figures 1 and 2). The basic underlying discussion was the definition of what the potter was thinking when he formed and deco-

Figure 1. Kent Flannery's drawing as a doctoral exam for Linda Manzanilla during the First Workshop of Advanced Training in Archaeology, 1973 (redrawn from Flannery's original).

Figure 2. Linda Manzanilla's response to the exam (redrawn from Manzanilla's original).

rated a vessel and whether the archaeologist could approach this issue. Flannery thought that a ceramic type was a group of coexisting forms with a particular decoration, an abstraction, and nobody could be sure what the potter was thinking when he decorated a vessel; the potter's perception of colors and forms obviously could be different from ours. The archaeologist may choose the decoration, the decoration together with the form, or the decoration together with the form and the function to construct a type. My position was that when manufacturing a vessel, the potter had in mind the different activities in which the vessel would be used; different forms and decorations should be related to diverse vessel functions. Functional types could be found repeatedly in archaeological contexts that revealed the particular activities for which they were used (see Childe, 1972:13–14); thus the archaeologist should create ceramic types that take into account the contexts in which the vessels appear, together with their form and decoration.

Although we did not agree on everything, Flannery's methodological perspective had a tremendous impact on my research. To conclude this postscript I outline that influence through a discussion of present and future perspectives on spatial research at Teotihuacan.

The study of activity areas and domestic structures has been a fertile field of research in archaeology. We consider the activity area as the minimal spatial unit of archaeological research, in which repeated social actions are reflected in particular concentrations and associations of raw materials, instruments, semiprocessed products, and debris, which relate to particular processes of production, use/consumption, storage, and disposal (Manzanilla, 1986a:11). The domestic group, in turn, may be defined by the individuals who share the same physical space for eating, sleeping, breeding, working, and resting; that is, they share residence, common activities, and kinship (Manzanilla, 1986a:14).

During the last 25 years, some Mexican scholars (Manzanilla, 1985a, 1987a, 1988–1989, 1990, 1993, 1996a; Manzanilla and Barba, 1990; Manzanilla et al., 1996; Reyna Robles, 1977; Serra Puche, 1980, 1988) have focused on this kind of research and believe in the enormous potential of extensive excavations for understanding artifact/raw materials/debris associations in discrete contexts. We have based our work on the premise that a domestic group is essentially one that shares activities (Ashmore and Wilk, 1988:3), and new methodological tools have been devised to reconstruct activity patterns in domestic compounds.

For the study of an early Mesoamerican city such as Teotihuacan, we propose the consideration and analysis of different scales to dissect social, economic, and ethnic variability through time. At the level of activity areas, we have designed an interdisciplinary methodology through the articulation and interpretation of distributional patterns in archaeological, osteological, chemical, paleobotanical, palinological, phytolithic, and paleozoological data in specific rooms of each apartment compound (Manzanilla, 1993, 1996a, 1988–1989; Manzanilla and Barba, 1990). These data are obtained in carefully described and sampled extensive excavations.

Natural and cultural transformation processes, as well as circumstances of abandonment, may move artifacts, raw materials, and debris away from the areas where they were used. Yet chemical compounds in stucco floors are trapped in a matrix that does not allow much modification (Barba, 1986; Barba et al., 1991), and thus activity areas may be determined even when few larger materials (artifacts) remain on top of the floors.

Our analysis of floor samples consists of three stages: 1) semiquantitative inorganic tests to define particular activity areas based on the analysis of the residues of chemical compounds in the soil (phosphate, calcium carbonate, pH); 2) quantitative inorganic techniques for greater accuracy; and 3) semiquantitative organic tests on particular substances, especially fatty acids, carbohydrates, and albumin (Barba et al., 1996:143). Chemical and paleobotanical analyses are directed to the bottoms of vessels to determine food content and particular preparation practices.

The comparison of botanical macrofossils, phytoliths, and pollen from each activity area allows greater precision in completing a list of plant remains in each apartment compound. Differences in food habits per family, domestic group, barrio group, and city sector may be assessed.

For particular burials in apartment compounds, osteological analysis has been directed to the determination of age and sex, physical characteristics, paleopathologies, paleodiet, mortality curves, systemic stress, and cultural practices (such as cranial deformation and teeth mutilation). Samples from different individuals are taken for DNA analysis (Vargas et al., 1998) to trace kinship and group identification and for barium-strontium isotopic analyses to detect paleodietary variation (Burton and Price, 1990; Manzanilla et al., in press) and migratory status determination (Price, Grupe, and Schroter, 1994; Price, Johnson et al., 1994).

We have applied our interdisciplinary methodology to several Teotihuacan apartment compounds, particularly one located in the northwestern part of the city. Three household territories could be discerned in an area of 550 m^2, each of 183 m^2, with a kitchen, storeroom, dormitory, ritual courtyards, service patios, and funerary sectors. Particular family preferences, as well as domestic group activities, were identified (Manzanilla, 1993). We could not distinguish between male and female activities, but we did determine that one domestic group specialized in the preparation and application of lime stucco.

We observed spatial patterning in the disposition of functional sectors, which extended beyond the framework of nuclear households. Thus, in general, storage zones were found to the west, areas for refuse to the south, funerary areas in the middle of the eastern sector (although exceptions exist), and neonate burials primarily on a north-south band in the eastern third of the compound.

Individual household sectors in the compound could be isolated either by considering circulation alleys or access points (Sanders, 1994:19 ff.) or by mapping the different food consumption loci for each household, as we did at Oztoyahualco 15B:N6W3 (Manzanilla, 1993, 1996a). One household in each compound seems to have been the most active in bonding the household group to the rest of the community. This fact is evidenced by the deity whose iconography is most prevalent in this household's domestic space, as well as in the relative abundance of foreign raw materials in this unit.

When comparing different excavated apartment compounds in the city, even when there is variation in size, we have found that the differences in access to botanical and faunal resources are very slight between the compounds. All apartment compounds had similar access to plant resources including maize (Nal-Tel Chapalote, Palomero Toluqueño, and Conical), amaranth, beans, squash (*Cucurbita pepo, C. maxima, C. moschata*, and *C. ficifolia*), hot peppers, tomato, *Chenopodium, Portulaca,* cactus, agave, Mexican hawthorn (*Crataegus mexicana*), and Mexican cherries (*Prunus capuli*) (González, 1993; Manzanilla, 1985a, 1993; McClung de Tapia, 1979, 1980:162–163; Storey, 1992:64). Important faunal resources included rabbits and hares, deer, duck, and fish, and in lesser degrees armadillo, squirrel, goose, quail, dove, turtle, and lizard (Sanders, 1994:31; Starbuck, 1975; Valadez Azúa, 1993; Valadez and Manzanilla, 1988).

There are differences not only in the specialized activities that occurred in household groups of different compounds but also in dominant activities of households, which suggest group and family specializations. Differences in the number of high-status products, particularly decorated ceramic tripods or mural paintings, and variability in the quality of the construction itself have been noted.

Many of the well-known three-temple complexes found throughout the northern part of the ancient city may be the centers of barrio or neighborhood groups, where cult and exchange activities took place for a number of the specialized corporate groups living in apartment compounds around them. Other types of wards not involving three-temple complexes may be distinguished in the southern part of the city (Cabrera Castro, 1996).

There also are foreign wards in Teotihuacan: the Oaxaca Barrio (Tlailotlacan) to the southwest (Rattray, 1993; Spence, 1989, 1992); the Merchants' Barrio to the east (Rattray, 1987, 1988), where circular adobe domestic structures with Maya polychrome pottery have been found; and perhaps also a small enclave from Michoacán on the western fringe of the city (Cabrera Castro, personal communication, 1993).

We have begun a program to test the migratory status of these foreign inhabitants, through DNA as well as barium-strontium isotope analyses. These data, together with the types of animals and plants consumed, the way the food is prepared, their funerary practices, as well as the particular constructive characteristics of their dwellings, may open new perspectives to ethnic behavior in the city.

Through this discussion I have tried to illustrate that the life of an early Mesoamerican city is being unveiled little by little, with the help of some creative new interdisciplinary approaches that are instrumental for understanding past societies. However, it is essential to recognize that these steps are always taken with the insightful teachings of Kent Flannery behind us and guiding us.

REFERENCES

Ashmore, W., and Wilk, R. R., 1988, Household and Community in the Mesoamerican Past. In *Household and Community in the Mesoamerican Past*, edited by W. Ashmore and R. R. Wilk, pp. 1–27. University of New Mexico Press, Albuquerque.

Barba, L. A., 1986, La química en el estudio de áreas de actividad. In *Unidades habitacionales mesoamericanas y sus áreas de actividad*, edited by L. Manzanilla, pp. 21–39. Arqueología, Serie Antropológica 76. Instituto de Investigaciones Antropológicas, Universidad Nacional Autónoma de México, Mexico City.

Barba, L., Rodríguez, R., and Córdova, J. L., 1991, *Manual de técnicas microquímicas de campo para la arqueología*. Instituto de Investigaciones Antropológicas, Universidad Nacional Autónoma de México, Mexico City.

Barba, L. A., Ortiz, A., Link, K. F., López Luján, L., and Lazos, L., 1996, Chemical Analysis of Residues in Floors and the Reconstruction of Ritual Activities at the Templo Mayor, Mexico. In *Archaeological Chemistry. Organic, Inorganic and Biochemical Analysis*, edited by M. V. Orna, pp. 139–156. ACS Symposium Series No. 625. American Chemical Society, Washington, D.C.

Burton, J. H., and Price, T. D., 1990, The Ratio of Barium to Strontium as a Paleodietary Indicator of Consumption of Marine Resources. *Journal of Archaeological Science* 17:547–557.

Cabrera Castro, R., 1996, Las excavaciones en La Ventilla. Un barrio teotihuacano. Sociedad Mexicana de Antropología. *Revista Mexicana de Estudios Antropológicos* XLII:5–30.

Childe, V. G., 1972, *Introducción a la arqueología*. Ediciones Ariel, Barcelona.

Flannery, K. V., 1965, The Ecology of Early Food Production in Mesopotamia. *Science* 147(3663):1247–1256.

Flannery, K. V., 1972, The Origins of the Village as a Settlement Type in Mesoamerica and the Near East. In *Man, Settlement and Urbanism*, edited by P. J. Ucko, G. W. Dimbleby, and R. Tringham, pp. 23–53. Gerald Duckworth, London.

Flannery, K. V., 1973, The Origins of Agriculture. *Annual Review of Anthropology* 2:271–310.

Flannery, K. V., 1976b, Research Strategy and Formative Mesoamerica. In *The Early Mesoamerican Village*, edited by K. V. Flannery, pp. 1–11. Academic Press, New York.

Flannery, K. V., 1982, Review of *In the Land of the Olmec*, Volumes 1 and 2, by M. D. Coe and R. A. Diehl. *American Anthropologist* 84:442–447.

Flannery, K. V., 1983, The Tierras Largas Phase and the Analytical Units of the Early Oaxacan Village. In *The Cloud People. Divergent Evolution of the Zapotec and Mixtec Civilizations*, edited by K. V. Flannery and J. Marcus, pp. 43–45. Academic Press, New York.

Flannery, K. V. (editor), 1976a, *The Early Mesoamerican Village*. Academic Press, New York.

Flannery, K. V., and Winter, M. C., 1976, Analyzing Household Activities. In *The Early Mesoamerican Village*, edited by K. V. Flannery, pp. 34–47. Academic Press, New York.

Flannery, K. V. (editor), 1986, *Guilá Naquitz. Archaic Foraging and Early Agriculture in Oaxaca, Mexico*. Academic Press, New York.

Flannery, K. V., and Marcus, J. (editors), 1983, *The Cloud People. Divergent Evolution of the Zapotec and Mixtec Civilizations*. Academic Press, New York.

González, J., 1993, A. Estudio del material arqueobotánico de Oztoyahualco. Capítulo 14. Macrofósiles botánicos, fitolitos y polen. In *Anatomía de un conjunto residencial teotihuacano en Oztoyahualco*, Volume 2, edited by L. Manzanilla, pp. 661–673. Universidad Nacional Autónoma de México, Mexico City.

Manzanilla, L., 1982, Hypothèses et indices du processus de formation de la civilisation égyptienne (cinquième et quatrième millénaires avant Jésus-Christ). Ph.D. dissertation, Université de Paris IV (Sorbonne), Paris.

Manzanilla, L., 1985a, El sitio de Cuanalan en el marco de las comunidades pre-urbanas del Valle de

Teotihuacan. In *Mesoamérica y el Centro de México*, edited by J. Monjarás-Ruiz, E. Pérez Rocha, and R. Brambila, pp. 133–178. Colección Biblioteca del INAH, Instituto Nacional de Antropología e Historia, Mexico CIty.

Manzanilla, L., 1985b, Le Développement des sociétés prédynastiques en Egypte: Considérations méthodologiques. In *Studi di paletnologia in Onore di Salvatore M. Puglisi*, edited by M. Liverani, A. Palmieri, and R. Peroni, pp. 485–494. Universitá di Roma La Sapienza, Rome.

Manzanilla, L., 1985c, Templo y palacio: proposiciones sobre el surgimiento de la sociedad urbana y el estado. *Anales de Antropología* 22:91–114. Universidad Nacional Autónoma de México, Mexico City.

Manzanilla, L., 1986a, Introducción. In *Unidades habitacionales mesoamericanas y sus áreas de actividad*, edited by L. Manzanilla, pp. 9–18. Arqueología, Serie Antropológica 76. Instituto de Investigaciones Antropológicas, Universidad Nacional Autónoma de México, Mexico City.

Manzanilla, L., 1986b, *La constitución de la sociedad urbana en Mesopotamia. Un proceso en la historia*. Serie Antropológica 80. Instituto de Investigaciones Antropológicas, Universidad Nacional Autónoma de México, Mexico City.

Manzanilla, L., 1986c, Cambios en la economía de subsistencia de los grupos prehistóricos del norte de Africa: El Nilo. *Anales de Antropología* 23:15–27. Universidad Nacional Autónoma de México, Mexico City.

Manzanilla, L., 1988–89, The Study of Room Function in a Residential Compound at Teotihuacan, Mexico. Universitá La Sapienza, Rome. *Origini, Giornate in onore di Salvatore Maria Puglisi* 14:175–186.

Manzanilla, L., 1990, Niveles de análisis en el estudio de unidades habitacionales. *Revista Española de Antropología Americana* 20:9–18.

Manzanilla, L., 1991, Pensamiento político y religión en Egipto y Mesopotamia durante el IV milenio a.C: Dos procesos de formación del estado. Universidad Nacional Autónoma de México, Mexico City, Mexico. *Anales de Antropología* 28:191–204.

Manzanilla, L., 1992a, *Akapana: una pirámide en el centro del mundo*. Instituto de Investigaciones Antropológicas, Universidad Nacional Autónoma de México, Mexico City.

Manzanilla, L., 1992b, The Economic Organization of the Teotihuacan Priesthood. In *Art, Ideology, and the City of Teotihuacan*, edited by J. Berlo, pp. 321–338. Dumbarton Oaks, Washington, D.C.

Manzanilla, L., 1995, 24. El altiplano central. In *Historia antigua de México*, Volume 2, edited by L. Manzanilla and L. López Luján, pp. 139-173. INAH-UNAM-Editorial Miguel Angel Porrúa, Mexico City.

Manzanilla, L., 1996a, Corporate Groups and Domestic Activities at Teotihuacan. *Latin American Antiquity* 7:245–266.

Manzanilla, L., 1996b, La organización económica de Teotihuacan y Tiwanaku. In *Administration in Ancient Societies*, edited by P. Ferioli, E. Fiandra, and G. G. Fissore, pp. 113–159. Centro Internazionale di Ricerche Archeologiche, Antropologiche e Storiche, Scriptorium, Turin.

Manzanilla, L., 1997, Teotihuacan: Urban Archetype, Cosmic Model. In *Emergence and Change in Early Urban Societies*, edited by L. Manzanilla, pp. 109–131. Plenum, New York.

Manzanilla, L., and Barba, L., 1990, The Study of Activities in Classic Households: Two Case Studies from Cobá and Teotihuacan. *Ancient Mesoamerica* 1:41–49.

Manzanilla, L., López, C., and Freter, A. C., 1996, Dating Results from Excavations in Quarry Tunnels behind the Pyramid of the Sun at Teotihuacan. *Ancient Mesoamerica* 7:245–266.

Manzanilla, L., and McClung de Tapia, E., 1997, Patrones de utilización de recursos durante las ocupaciones de túneles posteotihuacanos. *Cuicuilco* 10–11:107–120.

Manzanilla, L., Tejeda, S., and Martínez, J. C., 2000, Implicaciones del análisis de calcio, estroncio y zinc en el conocimiento de la dieta y la migración en Teotihuacan, México. *Anales de Antropología* 33:13–28. Universidad Nacional Autónoma de México, Mexico City.

Manzanilla, L. (editor), 1987a, *Cobá, Quintana Roo: Análisis de dos unidades habitacionales mayas del horizonte Clásico*. Arqueología, Serie Antropológica 82. Instituto de Investigaciones Antropológicas, Universidad Nacional Autónoma de México, Mexico City.

Manzanilla, L. (editor), 1987b, *Studies in the Neolithic and Urban Revolutions: The V. Gordon Childe Colloquium, Mexico 1986*. BAR International Series 349. British Archaeological Reports, Oxford.

Manzanilla, L. (editor), 1993. *Anatomía de un conjunto residencial teotihuacano en Oztoyahualco*. Instituto de Investigaciones Antropológicas, Universidad Nacional Autónoma de México, Mexico City, Mexico.

Marcus, J., and Flannery, K. V., 1996, *Zapotec Civilization: How Urban Society Evolved in Mexico's Oaxaca Valley*. Thames and Hudson, London.

McClung de Tapia, E., 1979, *Plants and Subsistence in the Teotihuacan Valley A.D. 100–750*. Ph.D. dissertation, Brandeis University, Boston. University Microfilms, Ann Arbor, Michigan.

McClung de Tapia, E., 1980, Interpretación de restos botánicos procedentes de sitios arqueológicos. *Anales de Antropología* 17:149–165. Universidad Nacional Autónoma de México, Mexico City.

Price, T. D., Grupe, G., and Schroter, P., 1994, Reconstruction of Migration Patterns in the Bell Beaker Period by Stable Strontium Isotope Analysis. *Applied Geochemistry* 9:413–417.

Price, T. D., Johnson, C. M., Ezzo, J. A., Ericson, J., and Burton, J. H., 1994, Residential Mobility in the Prehistoric Southwest United States. A Preliminary Study Using Strontium Isotope Analysis. *Journal of Archaeological Science* 21:315–330.

Rattray, E. C., 1987, Los barrios foráneos de Teotihuacan. In *Teotihuacan. Nuevos datos, nuevas síntesis y nuevos problemas*, edited by E. McClung de Tapia, and E. C. Rattray, pp. 243–273. Universidad Nacional Autónoma de México, Mexico City.

Rattray, E. C., 1988, Nuevas interpretaciones en torno al Barrio de los Comerciantes. *Anales de Antropología* 25:165–182. Universidad Nacional Autónoma de México, Mexico City.

Rattray, E. C., 1993, *The Oaxaca Barrio at Teotihuacan*. Monografías Mesoamericanas 1. Instituto de Estudios Avanzados, Universidad de las Américas-Puebla, Cholula.

Reyna Robles, R. M., 1977, Desarrollo y evolución de una "unidad habitacional" en una aldea Preclásica del altiplano central: Loma Torremote como un ejemplo. In *XV Mesa Redonda*, Volume I, pp. 357–383. Sociedad Mexicana de Antropología, Guanajuato.

Sanders, W. T., 1994, *The Teotihuacan Valley Project. Final Report. Volume 3. The Teotihuacan Occupation of the Valley, Part 1. The Excavations*. Occasional Papers in Anthropology No. 19. Pennsylvania State University, University Park.

Serra Puche, M. C., 1980, La unidad habitacional en Terremote-Tlaltelco, D. F: un análisis de actividad. Primera parte. *Anales de Antropología* 18:167–186. Universidad Nacional Autónoma de México, Mexico City.

Serra Puche, M. C., 1988, *Los recursos lacustres de la Cuenca de México durante el Formativo*. Colección Posgrado 3. Coordinación General de Estudios de Posgrado, Instituto de Investigaciones Antropológicas, Universidad Nacional Autónoma de México, Mexico City.

Spence, M. W., 1989, Excavaciones recientes en Tlailotlacan, el Barrio Oaxaqueño de Teotihuacan. *Arqueología* 5:81-104. Universidad Nacional Autónoma de México, Mexico City.

Spence, M. W., 1992, A Comparative Analysis of Ethnic Enclaves. Paper presented at the 57th annual Meeting of the Society for American Archaeology, Pittsburgh, PA.

Starbuck, D., 1975, Man-Animal Relationships in Pre-Columbian Central Mexico. Ph.D. dissertation, Department of Anthropology, Yale University, New Haven, CT.

Storey, R., 1992, *Life and Death in the Ancient City of Teotihuacan: A Modern Paleodemographic Synthesis*. University of Alabama Press, Tuscaloosa.

Valadez Azúa, R., 1993, Macrofósiles faunísticos. In *Anatomía de un conjunto residencial teotihuacano en Oztoyahualco*, Volume 2, edited by L. Manzanilla, pp. 729–825. Universidad Nacional Autónoma de México, Mexico City.

Valadez, R., and Manzanilla, L., 1988, Restos faunísticos y áreas de actividad en una unidad habitacional de la antigua ciudad de Teotihuacan. *Revista Mexicana de Estudios Antropológicos* 343(1):147–168.

Vargas, R., Manzanilla, L., Paredes, B., and Salazar, Z., 1998, Genetic Assessment of Epiclassic and Early Postclassic Populations in the Valley of Teotihuacan. Paper presented at the 63rd Annual Meeting of the Society for American Archaeology, Seattle, WA.

Comprehensive Bibliography of Kent V. Flannery

Books

1964 *The Middle Formative of the Tehuacán Valley: Its Pattern and Place in Mesoamerican Prehistory.* Ph.D. dissertation, microfilm, University of Chicago Library, Chicago.

1967 *Early Cultures and Human Ecology in South Coastal Guatemala* (with M. D. Coe). Smithsonian Contributions to Anthropology Vol. 3. Smithsonian Institution, Washington, D.C.

1969 *Prehistory and Human Ecology of the Deh Luran Plain* (with F. Hole and J. A. Neely). Memoirs No. 1. Museum of Anthropology, University of Michigan, Ann Arbor.

1970 *Prehistory of the Tehuacán Valley, Volume III: Ceramics of the Tehuacán Valley* (with R. S. MacNeish and F. A. Peterson). R. S. Peabody Foundation, Andover, MA, and University of Texas Press, Austin.

1976 *The Early Mesoamerican Village* (editor). Academic Press, New York.

1982 *Maya Subsistence: Studies in Memory of Dennis E. Puleston* (editor). Academic Press, New York.

1983 *The Cloud People: Divergent Evolution of the Zapotec and Mixtec Civilizations* (coeditor with J. Marcus). Academic Press, New York.

1986 *Guilá Naquitz: Archaic Foraging and Early Agriculture in Oaxaca, Mexico* (editor). Academic Press, New York.

1989 *The Flocks of the Wamani: A Study of Llama Herders on the Punas of Ayacucho, Peru* (with J. Marcus and R. G. Reynolds). Academic Press, New York.

1994 *Early Formative Pottery of the Valley of Oaxaca, Mexico* (with J. Marcus). Prehistory and Human Ecology of the Valley of Oaxaca, Volume 10. Memoirs No. 27. Museum of Anthropology, University of Michigan, Ann Arbor.

1996 *Zapotec Civilization: How Urban Society Evolved in Mexico's Oaxaca Valley* (with J. Marcus). Thames and Hudson, London and New York.

In prep. *Cueva Blanca: Archaic Hunters and White-Tailed Deer in Oaxaca, Mexico* (with F. Hole and R. G. Reynolds).

Articles

1961 Early Village Farming in Southwestern Asia. In *Patterns of Land Use and Other Papers*, edited by V. E. Garfield, pp. 7–17. Symposium Proceedings of the 1961 Annual Spring Meeting of the American Ethnological Society. University of Washington Press, Seattle.

1962 Excavations at Ali Kosh, Iran, 1961. *Iranica Antiqua* 2:97–148 (with F. Hole).

1964 Microenvironments and Mesoamerican Prehistory. *Science* 143:650–654 (with M. D. Coe).

1964 The Pre-Columbian Obsidian Industry of El Chayal, Guatemala. *American Antiquity* 30:43–49 (with M. D. Coe).

1965 Early Agriculture and Animal Husbandry in Deh Luran, Iran. *Current Anthropology* 6:105–106 (with F. Hole and J. Neely).

1965 The Ecology of Early Food Production in Mesopotamia. *Science* 147:1247–1256.

1966 The Postglacial "Readaptation" as Viewed from Mesoamerica. *American Antiquity* 31:800–805.

1966 En af Verden's Aeldste Landsbyer (The World's First Farmers). *Nationalmuseets Arbejdsmark*, 1966, pp. 85–96. Copenhagen, Danish National Museum (with P. Mortensen).

1966 Faunal Remains from the "Hut Sounding" at Eridu, Iraq. *Sumer* 23(1 and 2):61–63. Baghdad (with H. T. Wright).

1967 Animal Bones from Tell as-Sawan, Level III (Samarran Period). *Sumer* 23(1 and 2):179–182. Baghdad (with J. C. Wheeler).

1967 An Olmec "Were-Jaguar" from the Yucatán Peninsula. *American Antiquity* 32:109–111 (with G. Metcalf).

1967 Culture History versus Cultural Process: A Debate in American Archaeology. *Scientific American* 217(2):119–122.

1967 Farming Systems and Political Growth in Ancient Oaxaca. *Science* 158:445–454 (with A. V. T. Kirkby, M. J. Kirkby, and A. W. Williams, Jr.).

1967 The Prehistory of Southwestern Iran: A Preliminary Report. *Proceedings of the Prehistoric Society*, n.s. 33(9):147–206 (with F. Hole).

1967 Vertebrate Fauna and Hunting Patterns. In *The Prehistory of the Tehuacán Valley, Volume I: Environment and Subsistence*, edited by D. S. Byers, pp. 132–177. University of Texas Press, Austin.

1968 Archeological Systems Theory and Early Mesoamerica. In *Anthropological Archeology in the Americas*, edited by B. J. Meggers, pp. 67–87. Anthropological Society of Washington, Washington, D.C.

1968 The Olmec and the Valley of Oaxaca: A Model for Interregional Interaction in Formative Times. In *Dumbarton Oaks Conference on the Olmec*, edited by E. P. Benson, pp. 79–110. Dumbarton Oaks, Washington, D.C.

1968 Social and Economic Systems in Formative Mesoamerica. In *New Perspectives in Archeology*, edited by S. R. Binford and L. R. Binford, pp. 267–283. Aldine, Chicago (with M. D. Coe).

1969 Origins and Ecological Effects of Early Domestication in Iran and the Near East. In *The Domestication and Exploitation of Plants and Animals*, edited by P. J. Ucko and G. W. Dimbleby, pp. 73–100. Gerald Duckworth, London.

1969 The Faunal Remains from Sakheri Sughir. Appendix II in *The Administration of Rural Production in an Early Mesopotamian Town*, by H. T. Wright, pp. 143–149. Anthropological Papers No. 38. Museum of Anthropology, University of Michigan, Ann Arbor (with S. Bökönyi).

1969 An Analysis of Animal Bones from Chiapa de Corzo, Chiapas. Appendix II in *The Artifacts of Chiapa de Corzo, Chiapas, Mexico*, by T. A. Lee Jr., pp. 209–218. Anthropological Papers No. 26. New World Archaeological Foundation, Provo, UT.

1970 Climate and Man in Formative Oaxaca. *Archaeology* 23:144–152 (with J. Schoenwetter).

1972 Summary Comments: Evolutionary Trends in Social Exchange and Interaction. In *Social Exchange and Interaction*, edited by E. N. Wilmsen, pp. 129–135. Anthropological Papers No. 46. Museum of Anthropology, University of Michigan, Ann Arbor.

1972 The Origins of the Village as a Settlement Type in Mesoamerica and the Near East: A Comparative Study. In *Man, Settlement, and Urbanism*, edited by P. J. Ucko, R. Tringham, and G. W. Dimbleby, pp. 23–53. Gerald Duckworth, London.

1972 The Cultural Evolution of Civilizations. *Annual Review of Ecology and Systematics* 3:399–426.

1973 Archeology with a Capital "S." In *Research and Theory in Current Archeology*, edited by C. Redman, pp. 47–53. John Wiley, New York.

1973 The Origins of Agriculture. *Annual Review of Anthropology* 2:271–310.

1973 The Evolution of Technical Civilizations. In *Communication with Extraterrestrial Intelligence*, edited by C. Sagan, pp. 94–101. MIT Press, Cambridge, MA.

1975 La evolución cultural de las civilizaciones. *Cuadernos Anagrama, Serie: Sociología y Antropología* 103. Barcelona, Editorial Anagrama. (Spanish translation of 1972 *Annual Review of Ecology and Systematics* article.)

1975 La fauna de la región de Junín. *Revista del Museo Nacional* 41:77–78. Lima, Peru.

1976 Evolution of the Public Building in Formative Oaxaca. In *Cultural Change and Continuity*, edited by C. E. Cleland, pp. 205–221. Academic Press, New York (with J. Marcus).

1976 La evolución cultural de las civilizaciones. *Lecturas en aqueología, No. 2*, Museo de Arqueología y Etnología, Universidad Nacional Mayor de San Marcos. Lima, Peru. (Peruvian translation of 1972 *Annual Review of Ecology and Systematics* article, translated by Rosa Fung.)

1976 Formative Oaxaca and the Zapotec Cosmos. *American Scientist* 64:374–383 (with J. Marcus).

1976 The Nutritional Significance of Maize and Teosinte. *Ecology of Food and Nutrition* 4:243–249 (with J. R. K. Robson, R. I. Ford, and J. E. Konlande).

1978 Ethnoscience of the Sixteenth-Century Valley Zapotec. In *The Nature and Status of Ethnobotany*, edited by R. I. Ford, pp. 51–79. Anthropological Papers No. 67. ~~Museum of Anthropology, University of Michigan, Ann Arbor (with J. Marcus).~~

1981 The Preceramic and Formative of the Valley of Oaxaca. In *Supplement to the Handbook of Middle American Indians, Volume 1: Archaeology*, edited by J. A.

Sabloff, pp. 48–93. University of Texas Press, Austin (with J. Marcus and S. A. Kowalewski).

1982 The Golden Marshalltown: A Parable for the Archeology of the 1980s. *American Anthropologist* 84:265–278.

1982 Foreword. In *The Archaeology Workbook*, by S. Daniels and N. David, pp. xi–xii. University of Pennsylvania Press, Philadelphia.

1983 La Truelle d'or: parabole pour l'archeologie des années 1980. *Archaéologia* 178:43–53. (French translation of "The Golden Marshalltown," translated by Paul Courbin.)

1983 Early Pig Domestication in the Fertile Crescent: A Retrospective Look. In *The Hilly Flanks: Essays on the Prehistory of Southwestern Asia Presented to Robert J. Braidwood, November 15, 1982*, edited by T. C. Young Jr., P. E. L. Smith, and P. Mortensen, pp. 163–188. Studies in Ancient Oriental Civilization 36. Oriental Institute, University of Chicago, Chicago.

1985 Los orígenes de la agricultura en México: las teorías y la evidencia. In *Historia de la agricultura, época pehispánica—siglo XVI*, edited by T. Rojas Rabiela and W. T. Sanders, pp. 237–266. Colección Biblioteca del INAH, Serie: Historia. Instituto Nacional de Antropología e Historia, México.

1988 Comment on "Ecological Theory and Cultural Evolution in the Valley of Oaxaca" by W. T. Sanders and D. L. Nichols. *Current Anthropology* 29:57–58.

1989 Preceramic and Cave Occupations. In *Monte Albán's Hinterland, Part II*, by S. A. Kowalewski, G. M. Feinman, L. Finsten, R. E. Blanton, and L. M. Nicholas, pp. 39–53. Memoirs No. 23. Museum of Anthropology, University of Michigan, Ann Arbor (with L. Finsten and B. Macnider).

1990 Borrón y Cuenta Nueva: Setting Oaxaca's Archaeological Record Straight. In *Debating Oaxaca Archaeology*, edited by J. Marcus, pp. 17–69. Anthropological Papers No. 84. Museum of Anthropology, University of Michigan, Ann Arbor (with J. Marcus).

1990 Science and Science Fiction in Postclassic Oaxaca: Or "Yes, Virginia, There is a Monte Albán IV." In *Debating Oaxaca Archaeology*, edited by J. Marcus, pp. 191–205. Anthropological Papers No. 84. Museum of Anthropology, University of Michigan, Ann Arbor (with J. Marcus).

1993 Will The Real Model Please Stand Up: Comments on Saidel's "Round House or Square?" *Journal of Mediterranean Archaeology* 6:109–117.

1993 Cognitive Archaeology. *Cambridge Archaeological Journal* 3:260–270 (with J. Marcus). (Reprinted in 1996 in *Contemporary Archaeology in Theory*, edited by R. W. Preucel and I. Hodder, pp. 350–364. Blackwell, Oxford.)

1994 Ignacio Bernal, 1910–1992 (obituary). *American Antiquity* 59:72–76.

1994 Childe the Evolutionist: A Perspective from Nuclear America. In *The Archaeology of V. Gordon Childe: Contemporary Perspectives*, edited by D. Harris, pp. 101–119. Institute of Archaeology, University College London Press, Cambridge.

1994 Ancient Zapotec Ritual and Religion: An Application of the Direct Historical Approach. In *The Ancient Mind*, edited by C. Renfrew and E. B. W. Zubrow, pp. 55–74. Cambridge University Press, Cambridge (with J. Marcus).

1994 On The Perils of "Politically Correct" Archaeology. *Current Anthropology* 35:441–442 (with J. Marcus).

1994 Retired? You Must be Kidding! A Tribute to Ronald Spores. In *Caciques and Their People: A Volume in Honor of Ronald Spores*, edited by J. Marcus and J. F. Zeitlin, pp. 1–12. Anthropological Papers No. 89. Museum of Anthropology, University of Michigan, Ann Arbor (with J. Marcus).

1995 Prehistoric Social Evolution. In *Research Frontiers in Anthropology*, edited by C. R. Ember and M. Ember, pp. 1–26. Prentice-Hall, Englewood Cliffs, NJ.

1997 In Defense of the Tehuacán Project. *Current Anthropology* 38:660–662.

1998 The Ground Plans of Archaic States. In *Archaic States*, edited by G. M. Feinman and J. Marcus, pp. 15–57. School of American Research Press, Santa Fe.

1998 Introduction to the second edition of *The Archaeology of Society in the Holy Land*, edited by T. E. Levy, pp. xvii–xx. Leicester University, London.

In press Process and Agency in Early State Formation. *Cambridge Archaeological Journal* 8(2).

In press Ritual y religión de los antiguos zapotecos. In *Homenaje a José Luis Lorenzo*, edited by J. Litvak King and L. Mirambell. Universidad Autónoma de México, Mexico City (with J. Marcus).

In press Cultural Evolution in Oaxaca: The Rise of the Zapotec and Mixtec Civilizations. Chapter 8 in *Cambridge History of Native Peoples of the Americas*, edited by M. MacLeod and R. E. W. Adams. Cambridge University Press, Cambridge (with J. Marcus).

In press The Lower Pecos and the Mexican Plateau: A Comparison of Two Archaic Cultures. In *The Archaic of the Lower Pecos*, edited by H. J. Shafer, V. M. Bryant, and K. Sobolik. University of Texas Press, Austin.

In press The Present-Day Fauna of the Ayacucho Basin. In *The Prehistory of the Ayacucho Basin*, Volume 1, edited by R. S. MacNeish. University of Michigan Press, Ann Arbor (with E. S. Wing).

In press Animal Bones from Caves above 3,000 Meters. In *The Prehistory of the Ayacucho Basin*, Volume 1, edited by R. S. MacNeish. University of Michigan Press, Ann Arbor.

Book Reviews

1964 Review of A. E. Mourant and F. E. Zeuner, 1963, *Man and Cattle: Proceedings of a Symposium on Domestication*. *American Anthropologist* 66:944–947.

1965 Review of S. J. Olsen, 1964, *Mammal Remains from Archaeological Sites: Part 1, Southeastern and Southwestern United States*. *American Anthropologist* 67:1066–1067.

1965 Review of K. W. Butzer, 1965, *Environment and Archeology: An Introduction to Pleistocene Geography*. *American Journal of Archaeology* 69:372–373.

1965 Review of R. C. West (editor), 1964, *Handbook of Middle American Indians, Volume 1: Natural Environment and Early Cultures*. *American Anthropologist* 67:1333–1336.

1966 Review of A. Leeds and A. P. Vayda, 1965, *Man, Culture, and Animals: The Role of Animals in Human Ecological Adjustments*. *American Anthropologist* 68:1259–1262.

1967 Review of J. Paddock (editor), 1966, *Ancient Oaxaca: Discoveries in Mexican Archeology and History. American Journal of Archaeology* 71:427–429.

1969 Review of Alfonso Caso, Ignacio Bernal, and Jorge R. Acosta, 1967, *La cerámica de Monte Albán. American Journal of Archaeology* 73:104–107.

1977 Review of E. R. Wolf (editor), 1976, *The Valley of Mexico: Studies in Pre-hispanic Ecology and Society. Science* 196:759–761.

1977 Review of N. Hammond (editor), 1974, *Mesoamerican Archaeology: New Approaches. American Antiquity* 42:659–661.

1978 Review of H. Suzuki and F. Takai (editors), *The Paleolithic Site at Douara Cave in Syria: Parts I (1973) and II (1974). Bulletin of the American Schools of Oriental Research* 230:78–79.

1982 Review of M.D. Coe and R.A. Diehl, 1980, *In The Land of the Olmec*, Volumes 1 and 2. *American Anthropologist* 84:442–447.

1996 Review of F. Hours, O. Aurenche, J. Cauvin, M.-C. Cauvin, L. Copeland, and P. Sanlaville, 1994, *Atlas des sites du Proche Orient (14000–5700 B.P.). American Journal of Archaeology* 100:634.

Series Editor

"Prehistory and Human Ecology of the Valley of Oaxaca." *Memoirs*, Museum of Anthropology, University of Michigan. [Volumes 5–9, coeditor R. E. Blanton; Volume 10, coeditor J. Marcus.]

Vol. 1, 1973, *The Use of Land and Water Resources in the Past and Present Valley of Oaxaca, Mexico*, by A. V. T. Kirkby.

Vol. 2, 1973, *Sociopolitical Aspects of Canal Irrigation in the Valley of Oaxaca*, by S. H. Lees.

Vol. 3, 1975, *Formative Mesoamerican Exchange Networks with Special Reference to the Valley of Oaxaca*, by J. W. Pires-Ferreira.

Vol. 4, 1976, *Fábrica San José and Middle Formative Society in the Valley of Oaxaca*, by R. D. Drennan.

Vol. 5, 1978, Part 1: *The Vegetational History of the Oaxaca Valley*, by C. E. Smith Jr.; Part 2: *Zapotec Plant Knowledge*, by E. Messer.

Vol. 6, 1981, *Excavations at Santo Domingo Tomaltepec: Evolution of a Formative Community in the Valley of Oaxaca, Mexico*, by M. E. Whalen.

Vol. 7, 1982, *Monte Albán's Hinterland, Part 1*, by R. E. Blanton, S. A. Kowalewski, G. Feinman, and J. Appel.

Vol. 8, 1987, *Chipped Stone Tools in Formative Oaxaca, Mexico: Their Procurement, Production, and Use*, by W. J. Parry.

Vol. 9, 1989, *Agricultural Intensification and Prehistoric Health in the Valley of Oaxaca, Mexico*, by D. C. Hodges.

Vol. 10, 1994, *Early Formative Pottery of the Valley of Oaxaca, Mexico*, by K. V. Flannery and J. Marcus (with technical ceramic analysis by W. O. Payne).

The editors wish to express their gratitude to Joyce Marcus who compiled this bibliography.

Index

"*f*" or "*t*" following a page number indicates that the reader can find the term listed in a figure or table on the page.

Breinigsville, PA USA
06 August 2010
243079BV00004B/7/A

9 780306 462405